ILEANA TOMA
VALERICA MOȘNEGUȚU
ȘTEFANIA CONSTANTINESCU

SEQUENCES AND SERIES

AN INTRODUCTION, WITH APPLICATIONS AND EXERCISES

SERIES: MATHEMATICS FOR FUTURE ENGINEERS

VOLUME 1

Printed by CreateSpace

2018

Copyright © 2012 Ileana Toma
All rights reserved.

ISBN-13: 978-1986524063

ISBN-10: 198652406X

Preface

This book is addressed to all those who, after finishing the high school, wish a practical initiation in the domain of sequences and series.

This is the first volume of the series "Mathematics for future engineers". It has its origins in a course written for the first year students at the Technical University of Constructions of Bucharest.

To provide useful tools for (future) engineers and for specialists, in general, we put into evidence some practical applications of sequences and series (e.g., how to apply Lagrange's and Taylor's formula to the calculus of approximations, the catenary expressed in terms of hyperbolic functions, etc.).

We tried to make the involved mathematics as attractive as possible, by simplifying the presentation without loosing the mathematical rigor of the results. To increase accessibility and to encourage the reader to get a technical know-how about sequences and series, we provided for each newly introduced notion a series of applications and solved problems; each chapter ends by a section containing exercises and problems, each one of these being accompanied by hints and answers.

The sections marked by asterisks can be omitted, as well as some proofs. We introduced them, though, for the sake of a unitary and logical presentation.

The references contain, along with books, some links with sites which can be helpful for the reader.

We mention that this book is in no way endorsed or sponsored by CreateSpace, Amazon or their affiliates.

The Authors

CONTENTS

PREFACE 1

CHAPTER 1 5

   TOPOLOGY ON THE REAL AXIS 5

      1.1. Bounded sets 7

         1.1.1. Upper and lower bounds..9

         1.1.2. Infinite 11

      1.2. Limit points of a set 12

         1.2.1. Limit superior, limit inferior..14

      1.3. Countable sets 15

      Exercises 18

CHAPTER 2...19

   SEQUENCES AND SERIES OF REAL NUMBERS 19

      2.1. Sequences of real numbers..19

         2.1.1. The convergence of a sequence 20

         2.1.2. Operations with convergent sequences............................24

         2.1.3. Criteria of convergence for sequences.............................25

         2.1.4. Fundamental (Cauchy) sequence......................................27

      2.2. Series of numbers..31

         2.2.1. Convergence criteria for series with positive terms 37

         2.2.2. Some simple convergence criteria 40

         * 2.2.3. Other criterions of convergence 46

         2.2.4. Absolutely convergent series...51

         2.2.5. Series of complex numbers..57

      Exercises and problems...59

CHAPTER 3...64

   POWER SERIES, SEQUENCES AND SERIES OF FUNCTIONS 64

      3.1. Power series 64

         3.1.1. Domain of convergence, radius of convergence 65

         3.1.2. Operations with power series 73

3.2. Real functions, of one real variable...76
3.3. Lagrange's formula 82
    3.3.1. Forms of Lagrange's formula 83
    3.3.2. The geometrical interpretation of Lagrange's formula......................83
    3.3.3. Aplications of Lagrange's theorem 84
3.4. Taylor's formula 86
    3.4.1. The remainder of Taylor's formula 89
    3.4.2. Variants of Taylor's formula 90
    3.4.3. Taylor Polynomial 91
3.5. Power series expansions..94
3.6. Mac Laurin series for elementary functions...................................95
3.7. Sequences and series of functions..112
    3.7.1. The sequence-series equivalence.......................................115
    3.7.2. Uniform convergence, non-uniform convergence............................115
    3.7.3. Weierstrass criterion 117
    3.7.4. Properties of the sum of uniformly convergent series of functions..119
    3.7.5. Power series (Resumption)...122
    3.7.6. Computing limits of functions and definite integrals by means of power series 127
    3.7.7. Power series in complex 129
3.8. Fourier series..131
    3.8.1. Periodic functions 133
    3.8.2. Fourier series 134
    3.8.3. Fourier series of odd and even functions............................142
    3.8.4. Periodic extensions of functions.......................................146
Exercises and problems...150
REFERENCES 166

Chapter 1

TOPOLOGY ON THE REAL AXIS

We assume that the READER is FAMILIAR with the notions of sets and operations with sets, and we use the following notations:

$\mathfrak{N}$ — the set of **NATURAL** numbers,

$\mathfrak{Z}$ — the set of **INTEGER** numbers,

$\mathfrak{Q}$ — the set of **RATIONAL** numbers,

$\mathfrak{R}$ — the set of **REAL** numbers.

These sets are in the well-known strict inclusion relationships:

$$\mathfrak{N} \subset \mathfrak{Z} \subset \mathfrak{Q} \subset \mathfrak{R}.$$

It is also known that between the points of an oriented axis and the real numbers there is a one-to-one correspondence, represented by the figure below. This is why a set of real numbers $A \subset \mathfrak{R}$ can be also called a *linear set*.

A natural order relation can be introduced on the real axis: $a < b$ if a is situated on the left side of b.

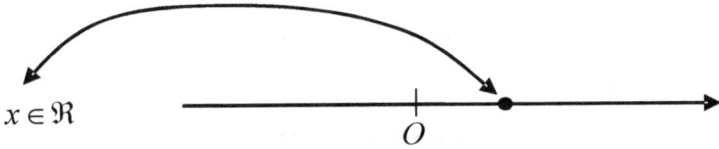

Figure 1.1. $\mathfrak{R}$ *is equivalent to the set of points of an oriented axis*

This equivalence allows us to introduce an order relation on $\Re$.

Further on, we will specify some notions familiar to the reader.

Definitions:

1. *Finite set*: A set which contains only a finite number of elements.

***The infinite* set** contains an infinite number of elements.

2. *The complement* of a set $A \subset \Re$: The set of the points of $\Re$ which do not belong to A, i.e.,

$$C_A = \{x \in \Re,\ x \notin A\}.$$

3. *Neighbourhood* of a point $a \in \Re$: A subset of $\Re$ situated at a smaller distance than a given $\varepsilon > 0$:

$$V_a = \{x \in \Re \mid |x - a| < \varepsilon\}.$$

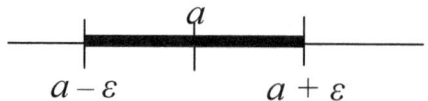

4. *The position of a point with respect to a set*

- $a \in \Re$ is an ***exterior point*** of A if one can find a neighbourhood V_a which does not contain any other point of A, except for a itself:

- $a \in \Re$ is an *interior point* of A if there exists a neighbourhood of a completely included in A:

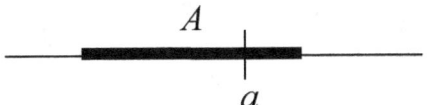

- $a \in \Re$ is a *boundary point* of A if any neighbourhood of a contains at least one point in A and at least one point in C_A:

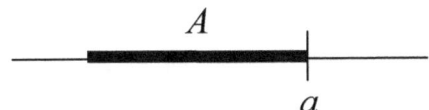

All these definitions can be correspondingly generalized in other topologies.

1.1. BOUNDED SETS

Definitions:

- $A \subset \Re$ is *bounded above* if one can find a number $b \in \Re$ such that $x \leq b, \ \forall \, x \in A$:

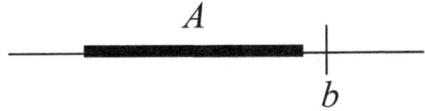

- $A \subset \Re$ is *bounded below* if one can find a number $a \in \Re$ such that $a \leq x, \ \forall \, x \in A$:

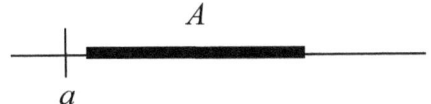

- $A \subset \Re$ is **bounded** if there exists $a, b \in \Re$ such that $a \leq x \leq b$, $\forall x \in A$ (or, equivalently, if one can find $c > 0$ such that $|x| \leq c$, $\forall x \in A$):

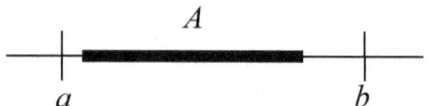

Examples:

- The set of real positive numbers $\Re_+$ is **bounded below**, because $x > 0$, $\forall x \in \Re_+$:

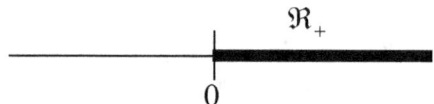

- The real negative numbers set $\Re_-$ is **bounded above**, because $x < 0$, $\forall x \in \Re_-$:

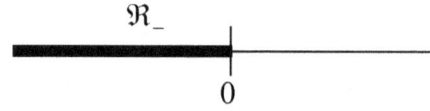

- The set $X = \{x \in \Re \mid x = \sin \alpha, \alpha \in \Re\}$ is **bounded**, because $|x| \leq 1$, $\forall x \in X$;

- The set $K = \left\{\dfrac{1}{n}, n \in \mathfrak{N}\right\}$ is **bounded**, because $0 < \dfrac{1}{n} \leq 1$ for any $n \in \mathfrak{N}$.

1.1.1. UPPER AND LOWER BOUNDS

Let $A \subset \Re$ bounded above. This means that there exists $b \in \Re$ such that $x \leq b$, $\forall x \in A$. The number b is called a **upper bound** of the set A. If $b' > b$, then, obviously, $x \leq b < b'$, $\forall x \in A$, hence b' is also a upper bound of A.

This means that a set of numbers bounded above has infinitely many upper bounds, which yields the natural problem: find "*the least upper bound*" of such a set:

Definition. *The least upper bound (LUB)* – also called *supremum* – of the set A bounded above is a number M with the following properties:

i) M is upper bound of A, meaning that $x \leq M$, $\forall x \in A$;

ii) for any $\varepsilon > 0$ we can find $x' \in A$ such that $x' > M - \varepsilon$.

We firstly prove that

1. M EXISTS!

* We can effectively build up the least upper bound of a set bounded above. Let L be the first integer bigger than any element of A. It follows that $L - 1$ is the integer part of M. We divide the interval $[L-1, L]$ into 10 equal parts. Let l_1 be the first division on the right side of A. Then $l_1 - 1$ is **the first decimal place** of M. By repeating the process, we find M, written in decimal form.

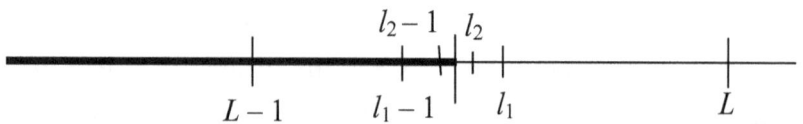

Figure 1.2. How to build up the upper bound of a set bounded above

2. M is UNIQUE!

* We assume, on the contrary, that there are two different numbers, M_1, M_2, both of them least upper bounds of A. Without restricting the generality, we can assume that $M_1 < M_2$. Let $\varepsilon = \dfrac{M_2 - M_1}{2}$. Then

- as M_2 is least upper bound of A, we can find, according to property ii), a number $x' \in A$ such that

$$x' > M_2 - \frac{M_2 - M_1}{2} = \frac{M_1 + M_2}{2};$$

- as M_1 is also least upper bound of A, it follows, according to property i), that $x' < M_1$.

Finally, we have

$$M_1 > \frac{M_1 + M_2}{2}, \text{ or } M_1 > M_2,$$

which is a contradiction. We thus proved that, for a set bounded above,

Theorem 1.1. *The least upper bound exists and it is unique.*

Similarly, we can introduce **the greatest lower bound** of a set $A \subset \Re$, bounded below:

Definion. *The greatest lower bound (GLB)* – also called *infimum* – of a set A bounded below is a number m with the following properties:

i) m is a lower bound of A, i.e., $x \geq m$, $\forall x \in A$;

ii) for any $\varepsilon > 0$ we can find $x' \in A$ such that $x' < m + \varepsilon$.

Following the same path as in theorem 1.1., we can prove that, for a set bounded below,

Theorem 1.2. *The greatest lower bound exists and it is unique.*

If $M, m \in A$, we say that the least upper bound/the greatest lower bound **is attained**.

Example. Let $A = \left\{ \dfrac{1}{2}, \dfrac{2}{3}, \ldots, \dfrac{n-1}{n}, \ldots \right\}$. We immediately see that $M = 1$, $m = \dfrac{1}{2}$. The least upper bound is **not attained**, because $1 \notin A$, but the greatest lower bound **is attained**, because $\dfrac{1}{2} \in A$.

1.1.2. INFINITE

The impropers "numbers" $\pm \infty$ – and also the work with them – are considered already known.

If an infinite set is *not* bounded above, we agree to say that it is **bounded above by** $+\infty$. Similarly, if an infinite set is *not* bounded below, we agree to say that it is **bounded below** by $-\infty$.

1.2. LIMIT POINTS OF A SET

Definitions
- α is called an ***accumulation point*** (or ***limit point***) of the set A if any of its neighbourhoods contains at least one point from A, different from α.
- An accumulation point of a set can be contained or not in that set.
- A set which contains its accumulation points is called ***closed***.

Example. The interval $[0, 1]$ is closed, but $[0, 1)$, $(0, 1]$, $(0, 1)$ are not closed.

- A set A is ***open*** if each point in A is an interior point.

Example. The interval $(0, 1)$ is open.

It can be proved that A is opened if and only if its complement C_A is closed.

There are infinite sets which do not have accumulation points (for example, $\mathcal{N}$, $\mathcal{Z}$). But

Theorem 1.3 (WEIERSTRASS-BOLZANO). *Every infinite, bounded set allows at least one accumulation point.*

* **Proof.** Let $A \subset \Re$ be infinite and bounded. It follows that we can find an interval $[a_0, b_0]$ which contains all the points of A, i.e., $A \subset [a_0, b_0]$. We divide this interval in two parts. At least one of these two new intervals contains infinitely many points of A. Let this interval be $[a_1, b_1]$. Obviously, $[a_1, b_1] \subset [a_0, b_0]$. We repeat the division and note with $[a_2, b_2] \subset [a_1, b_1] \subset [a_0, b_0]$ the interval containing infinitely many points of A (figure 1.3).

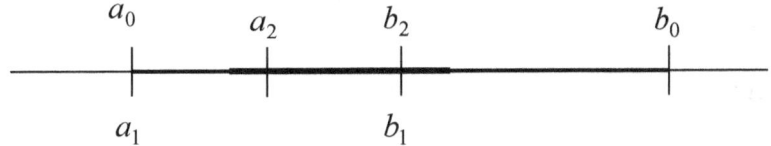

Figure 1.3. Dichotomy for the proof of Weierstrass-Bolzano's theorem

Finally, we obtain an infinite sequence of intervals $\{[a_n, b_n]\}_{n \in \Re}$, included one in each other, each of them containing infinitely many points of A. The length of the interval $[a_n, b_n]$ is

$$b_n - a_n = \frac{b_0 - a_0}{2^n},$$

wherefrom it results that $\{a_n\}, \{b_n\}$ are squeezing to a unique point α. This is, according to the definition, an accumulation point (or limit point) of A.

Remark. The limit point is unique in some cases, but this does not always occur. For example, by the above mentioned dichotomy we could have obtained, even from the first step, two intervals, each of them containing infinitely many points of A.

1.2.1. LIMIT SUPERIOR, LIMIT INFERIOR

Assume that A has several limit points. Let A^p be the set of these points. Representing these numbers on the real axis, and also keeping in mind the order relation on the real axis, it is naturally to think of "the smallest", and, accordingly, "the biggest" element of A^p, denoted respectively by l, L, of course, **if they exist !**

The number L is called the **limit superior** of A, and l – its **limit inferior**.

If A is bounded, then L and l have the following properties:

- for any $\varepsilon > 0$, on the left side of $l + \varepsilon$ there are infinitely many elements of A, and on the left side of $l - \varepsilon$ there are only a finite number;

- for any $\varepsilon > 0$, on the right side of $L - \varepsilon$ there are infinitely many elements of A, and on the right side of $L + \varepsilon$ there are only a finite number.

Figure 1.4. Properties of the limit points of a bounded set of reals

CONCLUSION: Given an infinite set $A \subset \Re$, we can define for it four points:

- the greatest lower bound m,
- the least upper bound M,
- the limit inferior l, and
- the limit superior L.

They obviously satisfy the following inequalities:
$$\boxed{m \leq l \leq L \leq M}.$$

Example.

Take $A = \left\{ x_n \in \Re \mid x_n = (-1)^n \dfrac{n^2+1}{n^2},\ n \in \mathfrak{N} \right\}$. In this case,

$$m = -2,\ M = \frac{5}{4},\ l = -1,\ L = 1.$$

1.3. COUNTABLE SETS

To count means to put in a one-to-one correspondence a finite set A of objects with a set of natural consecutive numbers, starting with 1.

What happens if A is infinite?

Definition. The set A is called ***countable*** if it can be put in a one-to-one correspondence with the naturals; more precisely, each of its elements receives an ordinal and two elements of different ordinals are different.

Example: $\mathfrak{N}$, $\mathfrak{Z}$, any sequence of real numbers $\{a_1, a_2, \ldots, a_n \ldots\}$.

The set $\mathbb{Q}$ of rational numbers is countable.

*Indeed, $\mathbb{Q}$ can be defined like this:

$$\mathbb{Q} = \left\{ \frac{p}{q} \,\middle|\, (p, q) = 1, q > 0 \right\}.$$

For each $m \in \mathfrak{N}$ we can consider, in ascending order following m, the fractions for which $m = |p| + q$. So, the rational numbers can be found uniquely in the following table

$m = 1:$ $\quad \dfrac{0}{1};$

$m = 2:$ $\quad -\dfrac{1}{1}; +\dfrac{1}{1};$

$m = 3:$ $\quad -\dfrac{2}{1}; -\dfrac{1}{2}; +\dfrac{1}{2}; +\dfrac{2}{1};$

$m = 4:$ $\quad -\dfrac{3}{1}; -\dfrac{1}{3}; +\dfrac{1}{3}; +\dfrac{3}{1};$

....

In this table, each rational number appears only once. Each line contains only a finite number of elements, so we can realize the correspondence with $\mathfrak{N}$.

The set $\mathfrak{R}$ of real numbers is NOT countable !

* It is enough to prove that the interval $I = (0, 1)$ is not countable. We suppose, by reductio ad absurdum, that I is

countable and let $\{a_1, a_2, \ldots, a_n, \ldots\}$ be its elements, which are numbers contained between 0 and 1. So, each a_j, $j \in \mathfrak{N}$, can be written as a decimal fraction as follows:

$$a_1 = 0.n_{11}n_{12}n_{13}\ldots\ldots\ldots n_{1k}\ldots\ldots$$
$$a_2 = 0.n_{21}n_{22}n_{23}\ldots\ldots\ldots n_{2k}\ldots\ldots$$
$$a_3 = 0.n_{31}n_{32}n_{33}\ldots\ldots\ldots n_{3k}\ldots\ldots$$
$$\ldots\ldots\ldots\ldots\ldots\ldots\ldots\ldots\ldots\ldots\ldots\ldots\ldots,$$

where $0 \leq n_{ij} \leq 9$.

Now, consider the number $\alpha = 0.n_1 n_2 n_3 \ldots\ldots\ldots n_k \ldots\ldots$, where $n_j \neq n_{jj}$. Obviously, $0 < \alpha < 1$, so $\alpha \in I$. But, by construction, α does not belong to the sequence $\{a_1, a_2, \ldots, a_n, \ldots\}$, because it is different from each element of the sequence by at least one decimal place. Consequently, I is not countable, hence even the less is $\mathfrak{R}$. ◻

Definition. Two sets A_1, A_2 are **equivalent** or **of the same cardinality** if they can be put in a one-to-one correspondence.

It results that a countable set is equivalent to $\mathfrak{N}$. The cardinal of $\mathfrak{N}$ is denoted by $\aleph_0$ (aleph zero).

The sets equivalent to $\mathfrak{R}$ are called **sets with cardinality of the continuum.** The cardinal of $\mathfrak{R}$ is denoted by c.

EXERCISES

Find the least upper bounds and the greatest lower bounds – as well as the limits superior and inferior – of the following sets:

1) $\left\{\dfrac{(-1)^n}{n}\right\}_{n \in \mathfrak{N}}$ A: $m = -1, M = 1, l = 0, L = 0$

2) $\left\{\dfrac{(-1)^n (n+1)}{2n}\right\}_{n \in \mathfrak{N}}$ A: $m = -1, M = \dfrac{3}{4}, l = -\dfrac{1}{2}, L = \dfrac{1}{2}$

3) The set Q of the rational numbers belonging to the interval $(0, 1)$. A: $m = 0, M = 1, l = 0, L = 1$

4) $\{\sin \pi / 2n\}_{n \in \mathfrak{N}}$ A: $m = 0, M = 1, l = 0, L = 0$

5) $\left\{\dfrac{n^2}{n+2}\right\}_{n \in \mathfrak{N}}$ A: $m = \dfrac{1}{3}, M = \infty, l = L = +\infty$

6) $\{(-1)^n n\}_{n \in \mathfrak{N}}$ A: $m = -\infty, M = \infty, l = -\infty, L = \infty$

7) $\left\{\dfrac{\sqrt{n}}{n+5}\right\}_{n \in \mathfrak{N}}$ A: $m = 0, M = \dfrac{1}{6}, l = L = 0$

8) $\left\{\dfrac{2n+3}{n}\right\}_{n \in \mathfrak{N}}$ A: $m = 2, M = 5, l = L = 2$

9) $\left\{1 - \dfrac{1}{n}\right\}_{n \in \mathfrak{N}}$ A: $m = 0, M = 1, l = L = 1$

10) $\left\{\dfrac{5^n - 1}{5^n + 1}\right\}_{n \in \mathfrak{N}}$ A: $m = \dfrac{2}{3}, M = 1, l = L = 1$

Chapter 2

SEQUENCES AND SERIES OF REAL NUMBERS

2.1. SEQUENCES OF REAL NUMBERS

Definition. A ***sequence of real numbers*** is a set of real numbers equivalent to $\mathcal{N}$.

Notations: $a_1, a_2, \ldots, a_n, \ldots$, or $\{a_n\}_{n \in \mathcal{N}}$;

a_n is ***the general term of the sequence.***

Example: For the sequence $\left\{\dfrac{1}{2}, \dfrac{4}{5}, \dfrac{9}{10}, \ldots\right\}$, the general term is $a_n = \dfrac{n^2}{n^2 + 1}$.

Let $\{a_n\}_{n \in \mathcal{N}}$ be a sequence of real numbers. As for any other set of real numbers, we can think of its least upper bound and its greatest lower bound, as well as of its limit superior L and its limit inferior l. For sequences, we generally use these notations:

$$L = \overline{\lim_{n \to \infty}} a_n, \quad l = \underline{\lim_{n \to \infty}} a_n.$$

Example: $\{a_n\}_{n \in \mathfrak{N}}$, where the general term is:

$$a_n = \frac{1}{2}\left[1 + (-1)^n\right].$$

Obviously, this sequence is a succession of 0 and 1: 0, 1, 0, 1, 0, 1,... .

This sequence does NOT coincide with the finite set $\{0, 1\}$; it is considered an infinite set, having two limit points: 0 and 1.

2.1.1. THE CONVERGENCE OF A SEQUENCE

Definition. A sequence $\{a_n\}_{n \in \mathfrak{N}}$ is **convergent** if there exists a real number a with the property that for any $\varepsilon > 0$ one can find an index N_ε such that

$$|a_n - a| < \varepsilon, \quad \forall n > N_\varepsilon. \tag{2.1.1}$$

If this inequality holds true, then a is called **the limit** of the sequence and we write

$$a = \lim_{n \to \infty} a_n. \tag{2.1.2}$$

Using the representation of numbers on the real axis, from the figure 2.1, it follows that

A sequence converges to $a \in \mathfrak{R}$ if there are only a finite number of its terms outside any neighborhood of a.

From the figure 2.1 it also follows that

i) on the left side of $a+\varepsilon$ we find infinitely many terms of the sequence, while on the left side of $a-\varepsilon$, only a finite number, so $a = \underline{\lim_{n \to \infty}} a_n$.

ii) on the right side of $a-\varepsilon$ there are infinitely many terms of the sequence, while on the right side of $a+\varepsilon$ we find only a finite number, so $a = \overline{\lim_{n \to \infty}} a_n$.

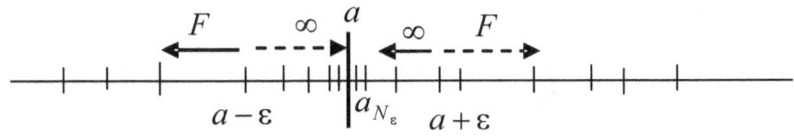

Figure 2.1. The limit of a sequence of real numbers

Examples

1. The sequence $\left\{1+(-1)^n\right\}_{n \in \mathfrak{N}} = \{0, 2, 0, 2, 0, 2, \ldots\}$ **IS NOT CONVERGENT,** because there are infinitely many terms of the sequence outside any neighborhood $(-\varepsilon, +\varepsilon)$ of 0, with $\varepsilon < 1$, more precisely $\{2, 2, 2, \ldots\}$. Also, there are infinitely many terms of the sequence outside any neighborhood of 2 of length smaller than 2, more precisely $\{0, 0, 0, \ldots\}$.

2. The sequence $\left\{\dfrac{1}{n}\right\}_{n \in \mathfrak{N}}$ **CONVERGES TO** 0, because any neighborhood of the origin contains almost all its terms, except for a finite number.

It follows that, if a sequence is convergent, then its limits superior and inferior coincide with the limit of the sequence. Moreover, from figure 2.1 we can see that if the limits superior and inferior of a sequence coincide with a, then a is the limit of the sequence.

Thus, it is proved that

Theorem 2.1. *A sequence* $\{a_n\}_{n \in \mathfrak{N}}$ *is convergent if and only if its limits superior and inferior coincide, i.e.,*

$$\overline{\lim_{n \to \infty}} a_n = \underline{\lim_{n \to \infty}} a_n. \qquad (2.1.3)$$

Examples

1. The sequence $\left\{\dfrac{1+(-1)^n}{2}\right\}_{n \in \mathfrak{N}} = \{0, 1, 0, 1, 0, 1, \ldots\ldots\}$

IS NOT CONVERGENT, because its limit superior, which is 1, does not coincide with its limit inferior, which is 0.

2. The sequence $\left\{\dfrac{1}{n}\right\}_{n \in \mathfrak{N}}$ **CONVERGES TO** 0. Indeed, both its limits superior and inferior are 0. By theorem 2.1, it follows that the sequence is convergent and its limit is 0.

Definition. *A divergent sequence* is a sequence $\{a_n\}_{n \in \mathfrak{N}}$ with the property that for any $M > 0$ there exists a index N_M such that

i) either $a_n > M$, $\forall n > N_M$,

ii) or $a_n < -M$, $\forall n > N_M$.

If i) is true, then we agree to write $\lim_{n \to \infty} a_n = +\infty$; in case ii), we have $\lim_{n \to \infty} a_n = -\infty$.

A sequence which is neither convergent, nor divergent, is called *oscillating*.

Remark. Usually, the denomination of divergent sequence is associated to a sequence which is not convergent; if we agree on this, then the definition of the divergent sequence will also cover the definition of the oscillating sequence.

From theorem 2.1 and from the definitions of divergent and oscillating sequences, it immediately follows that

- ♣ A non-constant convergent sequence has a unique accumulation point: its limit.

- ♣ A divergent sequence has also its limits superior and inferior equal, but infinite.

- ♣ A sequence is oscillating if its superior and inferior limits are distinct.

Theorem 2.2. *Every convergent sequence is bounded.*

* **Proof.** Suppose that the sequence $\{a_n\}_{n \in \mathfrak{N}}$ converges to a and let us consider $\varepsilon_0 > 0$ arbitrary, but fixed. Then, by the definition of convergence, there exists an index depending on ε_0 such that $|a_n - a| < \varepsilon_0$, $\forall n > N_{\varepsilon_0}$. Applying the properties of the modulus, it follows that

$$|a_n| < |a| + \varepsilon_0, \; \forall n > N_{\varepsilon_0}. \qquad (2.1.4)$$

Now, let $M = \max\left\{|a| + \varepsilon_0, |a_1|, |a_2|, \ldots, |a_{N_{\varepsilon_0}}|\right\}$. With this choice, we have

$$|a_n| < M_0, \; \forall n \in \mathfrak{N}, \qquad (2.1.5)$$

and the theorem is proved. ◻

2.1.2. OPERATIONS WITH CONVERGENT SEQUENCES

Consider the real and convergent sequences $\{a_n\}_{n \in \mathfrak{N}}$, $\{b_n\}_{n \in \mathfrak{N}}$ and let $a = \lim\limits_{n \to \infty} a_n$, $b = \lim\limits_{n \to \infty} b_n$. Then

♣ The sum/difference of the two sequences is convergent and

$$\lim_{n \to \infty}(a_n \pm b_n) = a \pm b = \lim_{n \to \infty} a_n \pm \lim_{n \to \infty} b_n. \qquad (2.1.6)$$

♣ The product of the two sequences is convergent and

$$\lim_{n \to \infty}(a_n \cdot b_n) = a \cdot b = \lim_{n \to \infty} a_n \cdot \lim_{n \to \infty} b_n. \qquad (2.1.7)$$

If $b \neq 0, b_n \neq 0, \forall n \in \mathfrak{N}$, then the sequence of the ratios $\left\{\dfrac{a_n}{b_n}\right\}_{n \in \mathfrak{N}}$ is convergent and

$$\lim_{n \to \infty}\left(\frac{a_n}{b_n}\right) = \frac{a}{b} = \frac{\lim\limits_{n \to \infty} a_n}{\lim\limits_{n \to \infty} b_n}. \qquad (2.1.8)$$

All these properties can be proved starting directly from the definition.

2.1.3. CRITERIA OF CONVERGENCE FOR SEQUENCES

The convergence or non-convergence of a sequence does not depend on a finite number of terms.

Definitions

1. A sequence $\{a_n\}_{n \in \mathfrak{N}}$ is called ***nondecreasing*** or ***monotonically increasing*** if

$$a_n \leq a_{n+1}, \quad \forall n \in \mathfrak{N}.$$

2. A sequence $\{a_n\}_{n \in \mathfrak{N}}$ is called ***nonincreasing*** or ***monotonically decreasing*** if

$$a_n \geq a_{n+1}, \quad \forall n \in \mathfrak{N}.$$

3. A sequence which is either monotonically increasing, or monotonically increasing is called ***monotonic***.

Theorem 2.3. *Any monotonic and bounded sequence is convergent.*

* **Proof.** If $\{a_n\}_{n \in \mathfrak{N}}$ is a constant sequence, then, obviously, it is convergent. Now, let the sequence $\{a_n\}_{n \in \mathfrak{N}}$ be nonconstant, monotonic and bounded. By Weierstrass-Bolzano's theorem (theorem 1.3), it follows that $\{a_n\}_{n \in \mathfrak{N}}$ allows at least one accumulation point. Suppose that this point is not unique, so the sequence would not be convergent. Then its limits superior and inferior would be distinct, i.e.,

$$\underline{\lim_{n \to \infty}} a_n \equiv l < L \equiv \overline{\lim_{n \to \infty}} a_n. \tag{2.1.9}$$

Let $\varepsilon = \dfrac{L-l}{3}$ and consider the intervals $I_1 = (l-\varepsilon, l+\varepsilon)$ and $I_2 = (L-\varepsilon, L+\varepsilon)$. These intervals are completely distinct and each of them contains infinitely many terms of the sequence. Let now $a_{n_1} \in I_1$. We can find an index $n_2 > n_1$ such that $a_{n_2} \in I_2$; otherwise, all the terms of the sequence, from the index n_1 on, would belong to I_1, which means that I_2 would not contain infinitely many terms. Thus, for $n_2 > n_1$ we have $a_{n_2} > a_{n_1}$. But, for the same reason as before, we can also find an index $n_3 > n_2$ such that $a_{n_3} \in I_1$. This means that $a_{n_2} > a_{n_3}$ (figure 2.2).

Consequently, the sequence cannot be monotonic, because

- for $n_2 > n_1$, $a_{n_2} > a_{n_1}$, and
- for $n_3 > n_2$, $a_{n_3} < a_{n_2}$.

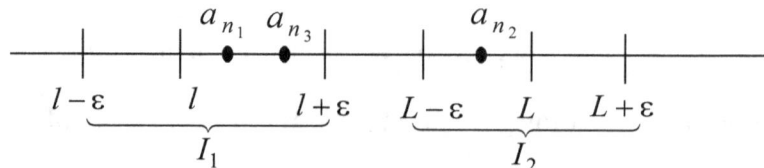

Figure 2.2. Proof of theorem 2.3

This contradicts the hypotheses, so the theorem is proved. ◻

2.1.4. FUNDAMENTAL (CAUCHY) SEQUENCE

Definition. A *fundamental sequence* or *Cauchy sequence* is a sequence $\{a_n\}_{n \in \mathfrak{N}}$ with the following property: for any $\varepsilon > 0$ there exists an index N_ε such that

$$|a_{n+p} - a_n| < \varepsilon, \ \forall n > N_\varepsilon, \ \forall p \in \mathfrak{N}. \tag{2.1.10}$$

Thus, if a sequence is Cauchy, then the bigger the index, the smaller the distance between the terms of the sequence.

Theorem 2.4. *Every Cauchy sequence is bounded.*

* **Proof.** Let $\{a_n\}_{n \in \mathfrak{N}}$ be a Cauchy sequence and let $\varepsilon_0 > 0$ be arbitrary, but fixed. As the sequence is fundamental, there exists an index $N_0 \equiv N(\varepsilon_0)$ such that

$$|a_{N_0+p} - a_{N_0}| < \varepsilon_0, \ \forall p \in \mathfrak{N}. \tag{2.1.11}$$

It follows that $|a_{N_0+p}| < |a_{N_0}| + \varepsilon_0, \ \forall p \in \mathfrak{N}$. If we note

$$M = \max\{|a_{N_0}| + \varepsilon_0, |a_1|, |a_2|, \ldots, |a_{N_0-1}|\}, \tag{2.1.12}$$

it immediately follows that $|a_n| < M, \ \forall n \in \mathfrak{N}$. ◻

Theorem 2.5 (THE FUNDAMENTAL THEOREM or CAUCHY'S THEOREM). *A sequence is convergent if and only if it is fundamental (Cauchy).*

* **Proof.** ⇒ *Necessity.* Let $\{a_n\}_{n \in \mathfrak{N}}$ be a sequence converging to a and let $\varepsilon > 0$. By definition, it follows that there exists an index N_ε from which on

$$|a_n - a| < \varepsilon, \quad \forall n > N_\varepsilon. \tag{2.1.13}$$

The more so,

$$|a_{n+p} - a| < \varepsilon, \quad \forall n > N_\varepsilon, \forall p \in \mathfrak{N}. \tag{2.1.14}$$

Applying the properties of the modulus, it follows that

$$\begin{gathered}|a_{n+p} - a_n| < |a_{n+p} - a| + |a_n - a| < 2\varepsilon, \\ \forall n > N_\varepsilon, \forall p \in \mathfrak{N},\end{gathered} \tag{2.1.15}$$

hence the sequence is fundamental.

⇐ *Sufficiency.* Let $\{a_n\}_{n \in \mathfrak{N}}$ be an non-constant Cauchy sequence. It is a bounded, infinite set, therefore, by Weierstrass-Bolzano's theorem (theorem 1.3), it allows at least one accumulation point. Suppose that this point is not unique; then, the sequence allows a limit superior L, distinct from its limit inferior l. Now, take $\varepsilon > 0$, $\varepsilon = \dfrac{L-l}{3}$. Then the intervals I_1, I_2 are completely distinct (figure 2.3). As l is a limit point of the sequence, we have $|a_{n_1} - l| < \varepsilon$, for any n_1 starting from some index N_1 on. Similarly, as L is also a limit point for $\{a_n\}_{n \in \mathfrak{N}}$, one must have $|a_{n_2} - L| < \varepsilon$, for any n_2 starting from some index N_2 on.

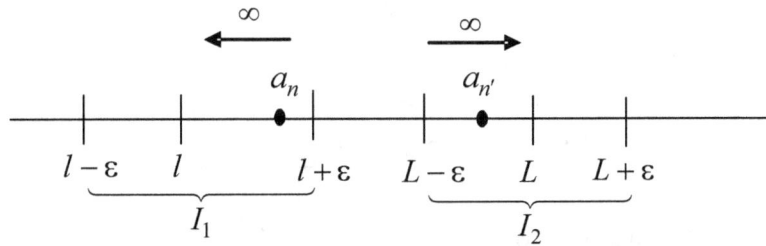

Figure 2.3. Proof of theorem 2.5

Now, let us take $N > \max\{N_1, N_2\}$. From the figure 2.3, we can see that there are infinitely many terms for which

$$|a_{n_2} - a_{n_1}| > \varepsilon = \frac{L-l}{3}, \quad \forall n_1, n_2 > N. \qquad (2.1.16)$$

This means that the sequence would not be fundamental, which contradicts the hypothesis. It follows that $L = l$ and, by applying theorem 2.1, we deduce that the sequence is convergent. ∎

Examples

1. The sequence $\{a_n\}_{n \in \mathfrak{N}}$, $a_n = 1 + \dfrac{1}{2} + \dfrac{1}{3} + \ldots + \dfrac{1}{n}$ **IS NOT CAUCHY.**

It is, obviously, monotonically increasing, and we have

$$a_{n+p} - a_n = \left(1 + \frac{1}{2} + \frac{1}{3} + \ldots + \frac{1}{n} + \frac{1}{n+1} + \ldots + \frac{1}{n+p}\right) -$$
$$-\left(1 + \frac{1}{2} + \frac{1}{3} + \ldots + \frac{1}{n}\right) = \frac{1}{n+1} + \frac{1}{n+2} \ldots + \frac{1}{n+p} > 0.$$

If $p > n$, then

$$|a_{n+p} - a_n| =$$
$$= \frac{1}{n+1} + \frac{1}{n+2} \ldots + \frac{1}{n+p} > \underbrace{\frac{1}{p+p} + \ldots + \frac{1}{p+p}}_{p \text{ times}} = \frac{1}{2}.$$

According to theorem 2.5, the sequence **IS NOT CONVERGENT**.

2. The sequence $\{a_n\}_{n \in \mathfrak{N}}$,

$$a_n = 1 - \frac{1}{2} + \frac{1}{3} + \ldots + (-1)^{n+1}\frac{1}{n},$$

IS CAUCHY.

Indeed,

$$|a_{n+p} - a_n| \le \frac{1}{n+1} - \frac{1}{n+2} \ldots + \frac{(-1)^{p-1}}{n+p} \le \frac{1}{n+1}, \quad \forall p \in \mathfrak{N}.$$

Let $\varepsilon > 0$. We use the inequality $\frac{1}{n+1} < \varepsilon$ to get N_ε.

Indeed, if $N_\varepsilon = \left[\frac{1-\varepsilon}{\varepsilon}\right]$, then it follows that $|a_{n+p} - a_n| < \varepsilon$, $\forall n > N_\varepsilon, \forall p \in \mathfrak{N}$, which means that the sequence is fundamental. According to theorem 2.5, it is convergent.

2.2. SERIES OF NUMBERS

Let $\{a_1, a_2, a_3, \ldots, a_n, \ldots\} \equiv \{a_n\}_{n \in \mathfrak{N}}$ be a sequence of real numbers.

The symbol

$$a_1 + a_2 + a_3 + \ldots + a_n + \ldots \equiv \sum_{n=1}^{\infty} a_n \qquad (2.2.1)$$

is called (infinite) **series**. The element a_n is **the general term** of the series.

Actually, we cannot calculate an infinite sum (this is why we used the term *symbol* to designate it), but we can add successively a_n, thus building a new sequence, $\{A_n\}_{n \in \mathfrak{N}}$:

$$\begin{aligned}
A_1 &= a_1, \\
A_2 &= a_1 + a_2, \\
A_3 &= a_1 + a_2 + a_3, \\
&\ldots\ldots\ldots\ldots\ldots\ldots\ldots\ldots \\
A_n &= a_1 + a_2 + \ldots + a_n, \\
&\ldots\ldots\ldots\ldots\ldots\ldots\ldots\ldots
\end{aligned} \qquad (2.2.2)$$

The elements of this sequence are called the **partial sums of the series**. We will always connect the series (2.2.1) with the sequence of its partial sums (2.2.2).

Definition. Let $A = \lim\limits_{n \to \infty} A_n$ be the finite – or infinite – limit of the sequence $\{A_n\}_{n \in \mathfrak{N}}$. This limit is called the **sum of the series** and we will write

$$A = a_1 + a_2 + a_3 + \ldots + a_n + \ldots = \sum_{n=1}^{\infty} a_n. \qquad (2.2.3)$$

With this definition, we can distinguish the nature of a series, taking into account the nature of the sequence of its partial sums. More precisely,

- ♣ the series is *convergent* if A exists and is finite;
- ♣ the series is *divergent* if $A = \pm\infty$;
- ♣ the series is *oscillating* if the sequence $\{A_n\}_{n \in \mathfrak{N}}$ has more than one accumulation point.

Examples

- **THE GEOMETRIC PROGRESSION** (frequently used) is expressed as

$$a + aq + aq^2 + \ldots + aq^n + \ldots . \qquad (2.2.4)$$

The number q is also called **ratio**. The general term of this series is $a_n = aq^n$.

Summing the first n terms, we find the partial sum:

$$A_n = a + aq + \ldots + aq^{n-1} = a\frac{1-q^n}{1-q}. \qquad (2.2.5)$$

We observe that

- ♣ if $|q| < 1$, then

$$\lim_{n \to \infty} A_n = \frac{a}{1-q} \lim_{n \to \infty} (1-q^n) = \frac{a}{1-q},$$

so the series is *convergent*;

♣ if $q > 1$, the sequence of the partial sum is divergent, so the series is ***divergent***;

♣ if $q = 1$, then $A_n = n$ and so, the series is ***divergent***;

♣ if $q = -1$, then the geometric progression becomes

$$a - a + a - a + a \ldots + (-1)^n a + \ldots .$$

The partial sums are, in this case

$$A_n = a - a + a - a + a \ldots + (-1)^{n-1} a = \begin{cases} 0 & n \text{ even} \\ a & n \text{ odd} \end{cases} ;$$

the sequence $\{A_n\}_{n \in \mathfrak{N}}$ has two limit points, 0 and a, so it is oscillating. Hence, the series is also ***oscillating***.

- **THE HARMONIC SERIES:**

$$1 + \frac{1}{2} + \frac{1}{3} + \ldots + \frac{1}{n} + \ldots . \qquad (2.2.6)$$

The general term is $a_n = \frac{1}{n}$, and its partial sum are

$$A_n = 1 + \frac{1}{2} + \frac{1}{3} + \ldots + \frac{1}{n}. \qquad (2.2.7)$$

We already analysed this sequence in the previous chapter and we proved that it is divergent.

Thus, we conclude that

The harmonic series is divergent.

Remark. The series (2.2.7) is called **HARMONIC** because each of its terms is the harmonic mean of the previous and the next term, i.e.,

$$\frac{2}{a_n} = \frac{1}{a_{n-1}} + \frac{1}{a_{n+1}}. \qquad (2.2.8)$$

Indeed,

$$\frac{2}{\frac{1}{n}} = \frac{1}{\frac{1}{n-1}} + \frac{1}{\frac{1}{n+1}}. \qquad (2.2.9)$$

- **THE ALTERNATING HARMONIC SERIES:**

$$1 - \frac{1}{2} + \frac{1}{3} - \ldots + (-1)^{n-1}\frac{1}{n} + \ldots \qquad (2.2.10)$$

was also previously analysed, by considering the sequence of its partial sums. We proved that this sequence is fundamental. Taking into account the fundamental theorem for sequences (theorem 2.5), we thus proved that

The alternating harmonic series is convergent.

Let us see how we can straightforwardly apply the fundamental theorem to a series, without resort to its partial sums.

From theorem 2.5, we infer that the series is convergent if and only if the sequence of its partial sums is Cauchy (or fundamental). This means that for any $\varepsilon > 0$ there exists an index N_ε such that

$$|A_{n+p} - A_n| < \varepsilon, \quad \forall n > N_\varepsilon, \forall p \in \mathfrak{N}. \qquad (2.2.11)$$

But

$$A_{n+p} - A_n = (a_1 + a_2 + \ldots + a_n + a_{n+1} + \ldots + a_{n+p}) - \\ -(a_1 + a_2 + \ldots + a_n) = a_{n+1} + a_{n+2} + \ldots + a_{n+p}. \qquad (2.2.12)$$

The sum

$$R_{np} \equiv a_{n+1} + a_{n+2} + \ldots + a_{n+p} \qquad (2.2.13)$$

is the so-called *finite remainder* of the series.

By applying directly the fundamental theorem for sequences (theorem 2.5), we can easily prove

Theorem 2.6 (THE FUNDAMENTAL THEOREM FOR SERIES). *The series* $\sum_{n=1}^{\infty} a_n$ *is convergent if and only if for any* $\varepsilon > 0$ *there exists an index* N_ε *such that the finite remainder* R_{np} *satisfies the inequality*

$$|R_{np}| < \varepsilon, \quad \forall n > N_\varepsilon, \forall p \in \mathfrak{N}. \qquad (2.2.14)$$

Now, let us neglect the first m terms of the series $\sum_{n=1}^{\infty} a_n$. We obtain

$$R_m \equiv a_{m+1} + a_{m+2} + \ldots + a_{m+p} + \ldots = \sum_{n=m+1}^{\infty} a_n, \qquad (2.2.15)$$

called the mth *order remainder* of the series. In its turn, R_m is also a series.

From all this, it follows that

Theorem 2.7. *If the series* $\sum_{n=1}^{\infty} a_n$ *is convergent, then* R_m *converges for any* $m \in \mathfrak{N}$. *Reciprocally, if there exists an order* m *for which* R_m *converges, then the series* $\sum_{n=1}^{\infty} a_n$ *is convergent.*

The **Proof** is imediate, using the Cauchy theorems for series and sequences. ◘

In theory, the theorems 2.6 and 2.7 can serve as criteria of convergence, but their application is difficult. However, we can prove

A NECESSARRY CONDITION OF CONVERGENCE:

Theorem 2.8. *If the series* $\sum_{n=1}^{\infty} a_n$ *is convergent, then* $\lim_{n \to \infty} a_n = 0$.

In other words,

The general term of a convergent series tends to **0.**

Proof. By theorem 2.6, the inequality (2.2.14) holds true for any natural p, hence also for $p = 1$. It follows that, for any $\varepsilon > 0$, we can find an index N_ε such that $|a_{n+1}| < \varepsilon$, $\forall n > N_\varepsilon$.

This means that $\lim_{n \to \infty} a_n = 0$ and the theorem is proved. ◘

ATTENTION !!! *The reciprocal of this theorem is not true!*

Example: The harmonic series

$$H \equiv 1 + \frac{1}{2} + \frac{1}{3} + \ldots + \frac{1}{n} + \ldots$$

is *divergent,* although *his general term tends to* 0!

2.2.1. CONVERGENCE CRITERIA FOR SERIES WITH POSITIVE TERMS

Consider the series with positive terms

$$a_1 + a_2 + a_3 + \ldots + a_n + \ldots \equiv \sum_{n=1}^{\infty} a_n, \quad a_n \geq 0, n \in \mathcal{N}. \quad (2.2.16)$$

We observe that its partial sums form always a monotonically increasing sequence. Indeed,

$$A_{n+1} = A_n + a_{n+1} \geq A_n.$$

Thus, *a series with positive terms is never oscillating.*

Theorem 2.9. *A series with positive terms is convergent if the sequence of its partial sum is bounded above; in this case, the sum of the series is precisely the least upper bound of the sequence. Otherwise, the series is divergent.*

Proof. It is imediately done by applying theorem 2.3. We showed that the sequence of the partial sums is monotonically increasing. As it is also bounded by hypothesis, it is convergent by theorem 2.3. ◘

A COMPARISON THEOREM

Theorem 2.10. Let $\sum_{n=1}^{\infty} a_n$, $\sum_{n=1}^{\infty} b_n$ *be two series with positive terms. Then*

i) if the series $\sum_{n=1}^{\infty} b_n$ converges and if there exists a constant $C > 0$ such that

$$a_n \leq C b_n, \ n \in \mathfrak{N}, \quad (2.2.17)$$

then the series $\sum_{n=1}^{\infty} a_n$ is convergent;

ii) if the series $\sum_{n=1}^{\infty} b_n$ diverges and if there exists o constant $C > 0$ such that

$$a_n \geq C b_n, \ n \in \mathfrak{N}, \quad (2.2.18)$$

then the series $\sum_{n=1}^{\infty} a_n$ is divergent.

Proof. Let us note by A_n, B_n the partial sums corresponding respectively to these series.

i) Writing the inequalities (2.2.17) successively and adding them member by member, we obtain

$$\begin{aligned} a_1 &\leq C b_1, \\ a_2 &\leq C b_2, \\ &\ldots\ldots\ldots\ldots \\ a_n &\leq C b_n, \\ A_n &\leq C B_n. \end{aligned} \quad (2.2.19)$$

As $\sum_{n=1}^{\infty} b_n$ converges, the sequence $\{B_n\}$ is bounded. Let B be its least upper bound. We have $A_n < CB$, hence the sequence $\{A_n\}$ is monotonically increasing and bounded.

By theorem 2.3, it results that $\{A_n\}$ is convergent.

ii) Suppose that the series $\sum_{n=1}^{\infty} a_n$ is convergent and let us write the inequalities (2.2.18) in the form

$$b_n \leq \frac{1}{C} a_n. \qquad (2.2.20)$$

Then, by the point *i)*, the series $\sum_{n=1}^{\infty} b_n$ is also convergent, which contradicts the hypothesis. Hence, the series $\sum_{n=1}^{\infty} a_n$ diverges. ◻

Remark. The above criteria can be applied even if the inequalities (2.2.17) and (2.2.18) are true except for a finite number of terms of the series.

Example. **THE RIEMANN SERIES:**

$$R(\alpha) \equiv 1 + \frac{1}{2^\alpha} + \frac{1}{3^\alpha} + \ldots + \frac{1}{n^\alpha} \ldots .$$

It is a series with positive terms, of general term $\frac{1}{n^\alpha}$.

a) **Case** $\alpha \leq 1$. We have $\frac{1}{n^\alpha} \geq \frac{1}{n}$. But $\frac{1}{n}$ is the general term of the harmonic series, which is divergent. So, in this case, according to the comparison theorem, point *ii)*, the Riemann series is also **divergent**.

b) **Case** $\alpha = 1$. Obviously, in this case, **the Riemann series coincides with the harmonic series**, which is **divergent**.

c) Case $\alpha > 1$. * For each $n \in \mathfrak{N}$, let us choose the first positive integer k for which $n < 2^k$. Then $A_n < A_{2^k-1}$. But A_{2^k-1} can be also written as

$$A_{2^k-1} = 1 + \left(\frac{1}{2^\alpha} + \frac{1}{3^\alpha}\right) + \left(\frac{1}{4^\alpha} + \frac{1}{5^\alpha} + \frac{1}{6^\alpha} + \frac{1}{7^\alpha}\right) + \ldots + \\ + \left(\frac{1}{2^{(k-1)\alpha}} + \frac{1}{2^{(k-1)\alpha}+1} + \ldots + \frac{1}{\left(2^k-1\right)^\alpha}\right), \quad (2.2.21)$$

hence

$$A_{2^k-1} \leq 1 + 2 \cdot \frac{1}{2^\alpha} + 4 \cdot \frac{1}{4^\alpha} + \ldots + 2^{k-1} \cdot \frac{1}{2^{(k-1)\alpha}},$$

$$A_{2^k-1} \leq 1 + \frac{1}{2^{\alpha-1}} + \frac{1}{2^{2(\alpha-1)}} + \ldots + 2^{k-1} \cdot \frac{1}{2^{(k-1)(\alpha-1)}} \leq \quad (2.2.22)$$

$$\leq \frac{1}{1 - \frac{1}{2^{\alpha-1}}}.$$

Finally, $A_n < \dfrac{1}{1 - \dfrac{1}{2^{\alpha-1}}}$, so the sequence of the partial sums of the series is bounded. Consequently, in this case the Riemann series is **convergent**.

2.2.2. SOME SIMPLE CONVERGENCE CRITERIA

Taking into account the comparison theorem on the one hand, and usual series, on the other hand, we can deduce some convergence criterions (tests) for series with positive terms, easily verified.

1. ROOT TEST (*Cauchy's test*). *A series with positive terms* $\sum_{n=1}^{\infty} a_n$ *is convergent if there exists a positive constant* $C < 1$ *such that the inequality*

$$\sqrt[n]{a_n} \leq C < 1, \quad n > N, \qquad (2.2.23)$$

holds for all of the terms of the series, from an index N on, and it is divergent if

$$\sqrt[n]{a_n} \geq C' > 1, \quad n > N', \qquad (2.2.24)$$

for some constant $C' > 1$, *from an index* N' *on.*

Proof. Indeed, if (2.2.23) is satisfied, then the inequality below is also valid:

$$a_n \leq C^n < 1, \quad n > N. \qquad (2.2.25)$$

But $\sum_{n=N}^{\infty} C^n$ is a geometric progression of subunitary ratio, hence it is convergent. By the theorem of comparison, part i), the series $\sum_{n=1}^{\infty} a_n$ converges too.

If the inequality (2.2.24) is valid, then

$$a_n \geq C'^n > 1, \quad n > N' \qquad (2.2.26)$$

and $\sum_{n=N'}^{\infty} C'^n$ is a geometric progression of ratio bigger than 1, hence it diverges. By the theorem of comparison, part *ii)*, the series $\sum_{n=1}^{\infty} a_n$ diverges too. ∎

Remark. In applications, it is simpler to take limits. We can easily prove that

♣ if $\lim_{n \to \infty} \sqrt[n]{a_n} < 1$, then the series is **convergent**, and

♣ if $\lim_{n \to \infty} \sqrt[n]{a_n} > 1$, then the series is **divergent**.

If $\lim_{n \to \infty} \sqrt[n]{a_n} = 1$, this criterion **DOES NOT WORK**. In this case, we must use a more refined criterion of convergence.

Example 1. Study the nature of the series

$$\frac{1}{3} + \left(\frac{2}{5}\right)^2 + \left(\frac{3}{7}\right)^2 + \ldots + \left(\frac{n}{2n+1}\right)^n + \ldots .$$

Solution. It is a numerical series with positive terms, of general term

$$a_n = \left(\frac{n}{2n+1}\right)^n.$$

Its form suggests the application of Cauchy's (root) criterion. Indeed, we have

$$\sqrt[n]{a_n} = \frac{n}{2n+1} < \frac{1}{2},$$

hence the inequality (2.2.23) is satisfied, for $C = \frac{1}{2} < 1$. The series is **convergent**.

Or, by using limits,

$$\lim_{n \to \infty} \sqrt[n]{a_n} = \lim_{n \to \infty} \frac{n}{2n+1} = \frac{1}{2} < 1.$$

Example 2. Study the nature of the series

$$\sum_{n=1}^{\infty} \frac{1}{2^n}\left(1 + \frac{1}{n}\right)^{n^2}.$$

Solution. We apply again the root criterion. The general term is

$$a_n = \frac{1}{2^n}\left(1 + \frac{1}{n}\right)^{n^2},$$

hence $\sqrt[n]{a_n} = \frac{1}{2}\left(1 + \frac{1}{n}\right)^n$, $n \in \mathfrak{N}$. We have

$$\lim_{n \to \infty} \sqrt[n]{a_n} = \lim_{n \to \infty} \frac{1}{2}\left(1 + \frac{1}{n}\right)^n = \frac{1}{2}e > 1,$$

therefore the series is **divergent**.

2. THE RATIO TEST (*D'Alembert's test*).

A series $\sum_{n=1}^{\infty} a_n$ *with positive terms is convergent if there exists a positive constant* $C < 1$ *such that the inequality*

$$\frac{a_{n+1}}{a_n} \leq C < 1, \quad n > N \tag{2.2.27}$$

holds true for all the terms of the series, from an index N on, and it is divergent if

$$\frac{a_{n+1}}{a_n} \geq C' > 1, \quad n > N', \tag{2.2.28}$$

for some constant $C' > 1$, from an index N' on.

Proof. Indeed, if (2.2.27) is satisfied, then the inequalities below are also valid

$$\begin{aligned} &a_2 \leq C a_1, \\ &a_3 \leq C a_2 \leq C^2 a_1, \\ &\dots\dots\dots\dots\dots\dots\dots\dots \\ &a_n \leq C^{n-1} a_1. \end{aligned} \tag{2.2.29}$$

But $\sum_{n=N}^{\infty} a_1 C^{n-1}$ is a geometric progression of subunitary ratio, hence it is convergent. By the first part of the theorem of comparison, the series $\sum_{n=1}^{\infty} a_n$ is convergent.

If inequality (2.2.28) holds, then

$$\begin{aligned} &a_2 \geq C' a_1, \\ &a_3 \geq C' a_2 \geq C'^2 a_1, \\ &\dots\dots\dots\dots\dots\dots\dots\dots \\ &a_n \geq C'^{n-1} a_1 \end{aligned} \tag{2.2.30}$$

and, as $a_1 \sum_{n=N'}^{\infty} a_1 C'^{n-1}$ is a geometric progression of ratio bigger than 1, it is divergent. According to part *ii)* of the comparison theorem, the series $\sum_{n=1}^{\infty} a_n$ is also divergent. ◘

Remark. For this test too, it is easier to take limits in applications. We can easily prove that

♣ if $\lim\limits_{n \to \infty} \dfrac{a_{n+1}}{a_n} < 1$, then the series is ***convergent***, and

♣ if $\lim\limits_{n \to \infty} \dfrac{a_{n+1}}{a_n} > 1$, then the series is ***divergent***.

If $\lim\limits_{n \to \infty} \dfrac{a_{n+1}}{a_n} = 1$, the criterion **CANNOT BE APPLIED**. We must apply a more refined convergence test.

Examples

1. Study the nature of the series

$$\frac{2}{1} + \frac{2^2}{2^{10}} + \frac{2^3}{3^{10}} + \ldots + \frac{2^n}{n^{10}} + \ldots .$$

Solution. It is a numerical series with positive terms, the general term being $a_n = \dfrac{2^n}{n^{10}}$. We apply the ratio test. We have

$$\lim_{n \to \infty} \frac{a_{n+1}}{a_n} = \lim_{n \to \infty} \frac{\dfrac{2^{n+1}}{(n+1)^{10}}}{\dfrac{2^n}{n^{10}}} = 2 \lim_{n \to \infty} \left(\frac{n}{n+1} \right)^{10} = 2 > 1,$$

therefore the series is divergent.

2. Study the nature of the series

$$\frac{1}{\sqrt{3}} + \frac{2}{3} + \frac{3}{3\sqrt{3}} + \ldots + \frac{n}{3^{\frac{n}{2}}} + \ldots .$$

Solution. We apply again the ratio test. The general term is $a_n = \dfrac{n}{3^{\frac{n}{2}}}$, hence

$$\lim_{n \to \infty} \frac{a_{n+1}}{a_n} = \lim_{n \to \infty} \frac{\frac{n+1}{3^{\frac{n+1}{2}}}}{\frac{n}{3^{\frac{n}{2}}}} = \lim_{n \to \infty} \frac{n+1}{n\sqrt{3}} = \frac{1}{\sqrt{3}} < 1;$$

thus, the series converges.

* 2.2.3. OTHER CRITERIONS OF CONVERGENCE

***1.* KUMMER'S CRITERION**: *The series with positive terms* $\sum_{n=1}^{\infty} a_n$ *is*

a) convergent, *if we can find a sequence of positive numbers* $\{\lambda_n\}_{n \in \mathfrak{N}}$, *such that from some N on the following inequality be true*

$$\lambda_n \frac{a_n}{a_{n+1}} - \lambda_{n+1} \geq \lambda > 0, \quad n > N; \quad (2.2.31)$$

b) *divergent*, *if there exists a sequence of positive numbers* $\{\lambda_n\}_{n\in\mathfrak{N}}$ *– the series* $\sum_{n=1}^{\infty}\dfrac{1}{\lambda_n}$ *being divergent – such that from some N on*

$$\lambda_n \frac{a_n}{a_{n+1}} - \lambda_{n+1} \leq \lambda < 0. \qquad (2.2.32)$$

The **Proof** uses the comparison theorem.

***PARTICULAR CASES*:**

I. $\lambda_n = 1,\ n \in \mathfrak{N}$. Then inequality *a)* becomes

$$\frac{a_n}{a_{n+1}} - 1 \geq \lambda > 0,\ n > N, \qquad (2.2.33)$$

which also reads:

$$\frac{a_n}{a_{n+1}} \geq \lambda + 1 > 1,\ n > N, \qquad (2.2.34)$$

or

$$\frac{a_{n+1}}{a_n} \leq \frac{1}{\lambda + 1} < 1,\ n > N, \qquad (2.2.35)$$

therefore we get the convergence condition from the ratio test. The inequality implying the divergence is similarly obtained from *b)*.

Hence, for $\lambda_n = 1,\ n \in \mathfrak{N}$, **Kummer's criterion coincides with the ratio criterion**.

II. $\lambda_n = n,\ n \in \mathfrak{N}$. Then inequality *a)* becomes

$$n\frac{a_n}{a_{n+1}} - (n+1) \geq \lambda > 0, \ n > N, \qquad (2.2.36)$$

it means that

$$n\left(\frac{a_n}{a_{n+1}} - 1\right) \geq \lambda + 1 > 0, \ n > N. \qquad (2.2.37)$$

We thus obtain

2. RAABE-DUHAMEL'S CRITERION: *The series with positive terms* $\sum_{n=1}^{\infty} a_n$ *is*

a) convergent, *if from an index N on, the inequality below is satisfied*

$$n\left(\frac{a_n}{a_{n+1}} - 1\right) \geq \lambda + 1 > 0, \ n > N, \qquad (2.2.38)$$

and it is

b) divergent, *if, from an index N on,*

$$n\left(\frac{a_n}{a_{n+1}} - 1\right) \leq \lambda + 1 < 0, \ n > N. \qquad (2.2.39)$$

Example. Consider the series with positive terms $\sum_{n=1}^{\infty} a_n$, for which the ratio of two consecutive terms is

$$\frac{a_{n+1}}{a_n} = \frac{n^\alpha + A_1 n^{\alpha-1} + A_2 n^{\alpha-2} + \ldots + A_\alpha}{n^\alpha + B_1 n^{\alpha-1} + B_1 n^{\alpha-2} + \ldots + B_\alpha}.$$

In this case, we see that the ratio test is useless, because

$$\lim_{n\to\infty}\frac{a_{n+1}}{a_n} = \lim_{n\to\infty}\frac{n^\alpha + A_1 n^{\alpha-1} + A_2 n^{\alpha-2} + \ldots + A_\alpha}{n^\alpha + B_1 n^{\alpha-1} + B_1 n^{\alpha-2} + \ldots + B_\alpha} = 1.$$

But, by applying Raabe-Duhamel's criterion, we get

$$\lim_{n\to\infty} n\left(\frac{a_n}{a_{n+1}} - 1\right) =$$

$$= \lim_{n\to\infty} n \frac{(B_1 - A_1)n^{\alpha-1} + (B_2 - A_2)n^{\alpha-2} + \ldots + (B_\alpha - A_\alpha)}{n^\alpha + B_1 n^{\alpha-1} + B_1 n^{\alpha-2} + \ldots + B_\alpha} =$$

$$= B_1 - A_1.$$

In conclusion:

- if $B_1 - A_1 > 1$, the series *converges*, and
- if $B_1 - A_1 < 1$ the series *diverges*.

Raabe-Duhamel's criterion is more refined than the ratio criterion.

3. THE LOGARITHMIC CRITERION: *The series with positive terms* $\sum_{n=1}^{\infty} a_n$ *is*

a) convergent, *if*

$$\frac{\log\frac{1}{a_n}}{\log n} \geq \lambda > 1, \quad n > N \tag{2.2.40}$$

and

b) divergent, *if*

$$\frac{\log\frac{1}{a_n}}{\log n} \leq \lambda < 1, \quad n > N. \tag{2.2.41}$$

The **Proof** uses the comparison theorem, with Riemann's series as term of comparison.

Sometimes, to get the sum of a convergent series, it could be more useful to rearrange its terms.

But is this new series convergent? If yes, has it the same sum?

Theorem 2.11. *If we rearrange the terms in a convergent series with positive terms, the sum of the series, as well as its nature, do not change.*

Proof. Let $a_1 + a_2 + \ldots + a_n + \ldots \equiv \sum_{n=1}^{\infty} a_n$ be a series with positive terms, of sum A.

By rearranging its terms, we obtain the series $a'_1 + a'_2 + \ldots + a'_n + \ldots \equiv \sum_{n=1}^{\infty} a'_n$. Obviously, among the terms of this last series we find, with other notations and otherwise placed, all the terms of the initial series.

Let $A'_n = a'_1 + a'_2 + \ldots + a'_n$ be the partial sum of order n of the modified series. By virtue of the previous *Remark*, we can find an index $N \geq n$ such that the inclusion $\{a'_1, a'_2, \ldots, a'_n\} \subseteq \{a_1, a_2, \ldots, a_N\}$ holds. Then, naturally, $A'_n \leq A_N$. But the series $\sum_{n=1}^{\infty} a_n$ converges to A, hence $A_N < A$. It means that $A'_n < A$ too. So, the sequence of the partial sums of the modified series is bounded above, and, as it is also monotonic, it follows that the series $\sum_{n=1}^{\infty} a'_n$ is convergent. Denoting its sum by A', it follows that

$$\boxed{A' \leq A}. \qquad (2.2.42)$$

We apply the same reasoning the other way around. For $A_n = a_1 + a_2 + \ldots + a_n$, we can find an index $N' \geq n$ such that the set $\{a_1, a_2, \ldots, a_n\} \subseteq \{a'_1, a'_2, \ldots, a'_{N'}\}$. So, it follows that $A_n \leq A'_{N'} < A'$, wherefrom

$$\boxed{A \leq A'}. \qquad (2.2.43)$$

The two inequalities (2.2.42), (2.2.43) involve $A' = A$. ◘

2.2.4. ABSOLUTELY CONVERGENT SERIES

The series with positive terms form a relatively restricted category, and the previous criteria can be applied only to this category. Yet, we can asociate to a series $\sum_{n=1}^{\infty} a_n$ its series of moduli (absolute values)

$$|a_1| + |a_2| + \ldots + |a_n| + \ldots \equiv \sum_{n=1}^{\infty} |a_n|, \qquad (2.2.44)$$

which is with positive terms. Thus, we can extend the possibilities of applications of the above criterions (tests).

Definition. A series $\sum_{n=1}^{\infty} a_n$ is called ***absolutely convergent*** if the series of its moduli $\sum_{n=1}^{\infty} |a_n|$ is convergent.

Following this definition, convergence is not in the least a compulsion for the series $\sum_{n=1}^{\infty} a_n$!

So, we must prove that

Theorem 2.12. *Every absolutely convergent series is also convergent.*

* **Proof.** We use the fundamental theorem for series. The finite remainder $R_{np} = \sum_{k=1}^{p} a_{n+k}$ always satisfies the inequality

$$\left| R_{np} \right| = \left| \sum_{k=1}^{p} a_{n+k} \right| \le \sum_{k=1}^{p} \left| a_{n+k} \right| \equiv \overline{R}_{n+p}, \qquad (2.2.45)$$

where we denoted the finite rest of the series of moduli by $\overline{R}_{n+p}$. Now, let $\varepsilon > 0$ be arbitrary, but fixed. By hypothesis, the series of moduli is convergent, hence there exists an index N_ε from which on

$$\overline{R}_{n+p} < \varepsilon, \quad \forall p \in \mathfrak{N}. \qquad (2.2.46)$$

According to inequality (2.2.46), it results that, the more so,

$$\left| R_{np} \right| < \varepsilon, \quad \forall n > N_\varepsilon, \forall p \in \mathfrak{N}. \qquad (2.2.47)$$

Inequality (2.2.47) involves the convergence of the series $\sum_{n=1}^{\infty} a_n$, according to implication $\Leftarrow$ from the fundamental theorem 2.6. ∎

Remark. The reciprocal of the theorem 2.12 is not generally valid. Indeed, the alternating series

$$1 - \frac{1}{2} + \frac{1}{3} - \frac{1}{4} + \ldots + (-1)^{n-1}\frac{1}{n} + \ldots$$

is, as we have seen, convergent, but its series of moduli

$$1 + \frac{1}{2} + \frac{1}{3} + \frac{1}{4} + \ldots + \frac{1}{n} + \ldots$$

is precisely the harmonic series, which is not convergent.

Definition. A convergent series which is not absolutely convergent is called *semi-convergent* (or *conditionally convergent*).

Thus,

The alternating series is semi-convergent.

* A SPECIAL CRITERION OF CONVERGENCE:

ABEL-DIRICHLET'S CRITERION

Theorem 2.13. *Let* $\sum_{n=1}^{\infty} a_n$ *be an arbitrary series and* $\{\lambda_n\}_{n \in \mathfrak{N}}$ *be a sequence of positive numbers, monotonically decreasing and convergent to* 0. *Then, if the sequence of its partial sums is bounded, i.e.,*

$$|A_n| \leq M, \quad n \in \mathfrak{N}, \qquad (2.2.48)$$

the series $\sum_{n=1}^{\infty} \lambda_n a_n$ *is convergent.*

* **Proof.** As $a_n = A_n - A_{n-1}$, we can write the finite remainder $R'_{np} \equiv \sum_{k=1}^{p} \lambda_{n+k} a_{n+k}$ of the series to be studied as

$$R'_{np} = \sum_{k=1}^{p} \lambda_{n+k} \left(A_{n+k} - A_{n+k-1} \right). \qquad (2.2.49)$$

By writing explicitly the sum from the right member, we get

$$\begin{aligned} R'_{np} &= \left(\lambda_{n+p} A_{n+p} - \lambda_{n+p} A_{n+p-1} \right) + \\ &+ \left(\lambda_{n+p-1} A_{n+p-1} - \lambda_{n+p-1} A_{n+p-2} \right) + \\ &+ \left(\lambda_{n+p-2} A_{n+p-2} - \lambda_{n+p-2} A_{n+p-3} \right) + \dots \\ &+ \left(\lambda_{n+1} A_{n+1} - \lambda_{n+1} A_n \right). \end{aligned} \qquad (2.2.50)$$

Grouping otherwise the terms, we have

$$R'_{np} = \lambda_{n+p} A_{n+p} - \lambda_{n+1} A_n + \sum_{k=n+1}^{n+p-1} \left(\lambda_k - \lambda_{k+1} \right) A_k, \qquad (2.2.51)$$

wherefrom, by applying modulus,

$$\begin{aligned} \left| R'_{np} \right| &\leq M \left(\lambda_{n+p} + \lambda_{n+1} \right) + M \sum_{k=n+1}^{n+p-1} \left(\lambda_k - \lambda_{k+1} \right) = \\ &= M \left(\lambda_{n+p} + \lambda_{n+1} \right) + \\ &+ M \left[\left(\lambda_{n+1} - \lambda_{n+2} \right) + \left(\lambda_{n+2} - \lambda_{n+3} \right) + \dots \right. \\ &\left. + \left(\lambda_{n+p-1} - \lambda_{n+p} \right) \right]. \end{aligned} \qquad (2.2.52)$$

As the sequence $\{\lambda_n\}_{n \in \mathfrak{N}}$ is monotonically decreasing, the brackets above contain positive differences; this is why we could get rid of moduli. Reducing like terms, we finally obtain the inequality

$$\left|R'_{np}\right| \leq 2M\lambda_{n+1}. \tag{2.2.53}$$

As $\lim\limits_{n \to \infty} \lambda_n = 0$, for any $\varepsilon > 0$ we can find an index N_ε such that $\lambda_{n+1} < \dfrac{\varepsilon}{2M}$, $\forall n > N_\varepsilon$.

Thus, $\left|R'_{np}\right| < \varepsilon$, $\forall n > N_\varepsilon, \forall p \in \mathfrak{N}$, such that the finite remainder of the series can be made arbitrarily small. By the fundamental theorem, it follows that the series $\sum\limits_{n=1}^{\infty} \lambda_n a_n$ is convergent. ◘

PARTICULAR CASE:
LEIBNIZ'S CRITERION FOR ALTERNATING SERIES

Definition. An *alternating series* is a series whose terms are alternately positive and negative.

An alternating series can be written as $\sum\limits_{n=1}^{\infty}(-1)^{n-1} b_n$, where $b_n > 0$, $\forall n \in \mathfrak{N}$.

Theorem 2.14 (LEIBNIZ). *If the sequence $\{b_n\}_{n \in \mathfrak{N}}$ of positive numbers is monotonically decreasing and $\lim\limits_{n \to \infty} b_n = 0$, then the alternating series $\sum\limits_{n=1}^{\infty}(-1)^{n-1} b_n$ is convergent.*

* **Proof.** We apply Abel-Dirichlet's criterion for $\lambda_n = b_n$ and $a_n = (-1)^{n-1}$. We must check if the hyppotheses of the theorem 2.13 are satisfied. By the choice we made, the sequence $\{|b_n|\}_{n \in \mathfrak{N}}$ is monotonically decreasing and tends to 0. Let us prove that the sequence of the partial sums of the series $\sum_{n=1}^{\infty} a_n \equiv \sum_{n=1}^{\infty} (-1)^{n-1}$ is bounded. We have

$$\begin{aligned} A_1 &= a_1 = 1, \\ A_2 &= A_1 + a_2 = 1 - 1 = 0, \\ A_3 &= A_2 + a_3 = 0 + 1 = 1, \end{aligned} \qquad (2.2.54)$$

$$\dots\dots\dots\dots\dots\dots\dots\dots\dots\dots\dots\dots$$

hence $|A_n| \leq 1$, $\forall n \in \mathfrak{N}$. The hypotheses of the theorem 2.13 are thus obviously satisfied; it results that the alternating series is convergent. ◘

Example. The alternating harmonic series

$$1 - \frac{1}{2} + \frac{1}{3} - \frac{1}{4} + \ldots + (-1)^{n+1} \frac{1}{n} + \ldots$$

is **convergent**, because the sequence of the moduli, i.e., $\left\{\frac{1}{n}\right\}_{n \in \mathfrak{N}}$, is monotonically decreasing and tends to 0.

IMPORTANT REMARK:

If the sequence of the moduli tends to 0, but it is not monotonically decreasing, it may occur that the alternating series not be convergent!

Example. Study the nature of the series

$$\frac{1}{\sqrt{2}-1} - \frac{1}{\sqrt{2}+1} + \frac{1}{\sqrt{3}-1} - \frac{1}{\sqrt{3}+1} +$$
$$+ \frac{1}{\sqrt{4}-1} - \frac{1}{\sqrt{4}+1} + \ldots + \frac{1}{\sqrt{n}-1} - \frac{1}{\sqrt{n}+1} + \ldots.$$

Solution. The series is alternating, but the sequence of the moduli

$$\frac{1}{\sqrt{2}-1}, \frac{1}{\sqrt{2}+1}, \frac{1}{\sqrt{3}-1}, \frac{1}{\sqrt{3}+1}, \ldots, \frac{1}{\sqrt{n}-1}, \frac{1}{\sqrt{n}+1}, \ldots,$$

while tending to 0, is ***not monotonically increasing***. Indeed,

$$\frac{1}{\sqrt{n}-1} > \frac{1}{\sqrt{n}+1}, \text{ but } \frac{1}{\sqrt{n}+1} < \frac{1}{\sqrt{n+1}-1}.$$

Besides, computing A_{2n}, we find

$$A_{2n} = \sum_{k=1}^{n} \left(\frac{1}{\sqrt{k+1}-1} - \frac{1}{\sqrt{k+1}+1} \right) = 2 \sum_{k=1}^{n} \frac{1}{k},$$

i.e, the double of the partial sum of the harmonic series, which is not convergent.

2.2.5. SERIES OF COMPLEX NUMBERS

The series

$$\sum_{n=1}^{\infty} a_n,$$

where

$$a_n = \alpha_n + i\beta_n,$$

is a series of complex numbers.

By definition, a series of complex numbers is convergent if each of the real numerical series

$$\sum_{n=1}^{\infty} \alpha_n, \quad \sum_{n=1}^{\infty} \beta_n$$

is convergent, and the sum of series is

$$A + iB, \text{ if } A = \sum_{n=1}^{\infty} \alpha_n \text{ and } B = \sum_{n=1}^{\infty} \beta_n.$$

A series of complex numbers is **absolutely convergent** if each of the series $\sum_{n=1}^{\infty} |\alpha_n|$ and $\sum_{n=1}^{\infty} |\beta_n|$ is convergent.

Theorem 2.15. *The series* $\sum_{n=1}^{\infty} (\alpha_n + i\beta_n)$ *is absolutely convergent if and only if the series of positive numbers* $\sum_{n=1}^{\infty} (\alpha_n^2 + \beta_n^2)$ *is convergent.*

Proof. If the series $\sum_{n=1}^{\infty} (\alpha_n^2 + \beta_n^2)$ is convergent, then, as

$$|\alpha_n| \leq \sqrt{\alpha_n^2 + \beta_n^2}, \quad |\beta_n| \leq \sqrt{\alpha_n^2 + \beta_n^2},$$

the series $\sum_{n=1}^{\infty} |\alpha_n|$ and $\sum_{n=1}^{\infty} |\beta_n|$ are also convergent, by the comparison theorem. Reciprocally, if $\sum_{n=1}^{\infty} |\alpha_n|$ and $\sum_{n=1}^{\infty} |\beta_n|$ are convergent, then, as the following inequalities

$$\sqrt{\alpha_n^2 + \beta_n^2} \leq |\alpha_n| + |\beta_n|,$$

are valid, the series $\sum_{n=1}^{\infty}\left(\alpha_n^2+\beta_n^2\right)$ is also convergent, by the same comparison theorem.

EXERCISES AND PROBLEMS

1. Knowing that $a+b+c=0$, find
$$\lim_{n\to\infty}\left(a\sqrt{n-2}+b\sqrt{n+5}+c\sqrt{n+1}\right).$$

A: 0

2. Calculate $\lim_{n\to\infty}\left(\dfrac{3n-2}{3n-5}\right)^n$.

A: e

3. Calculate $\lim_{n\to\infty}\left(\dfrac{4n^3+3n^2+5n-7}{4n^3+3n^2-5n+7}\right)^n$.

A: $\sqrt{e}$

4. Study the nature of the sequence $\{a_n\}_{n\in\mathfrak{N}}$, of general term
$$a_n=\frac{1}{n+1}+\frac{1}{n+2}+\ldots+\frac{1}{n+n}.$$

Hint: We prove that the sequence is monotonically increasing and bounded, hence it is convergent.

5. Consider the sequence $\left\{\dfrac{1}{n}\right\}_{n\in\mathfrak{N}}$. Verify if it is a Cauchy sequence (fundamental) and, in the affirmative, find its limit.

6. The same problem for the sequence $\left\{\dfrac{n}{n+1}\right\}_{n\in\mathfrak{N}}$.

7. Find $a \neq 0$ such that

$$l \equiv \lim_{n\to\infty} \dfrac{\sqrt{(2-a)^2 n^6 + 2n^3 + \sqrt{3}}}{an^3} = 3.$$

A: $a = \dfrac{1}{2}$

8. Find $\alpha, \beta \in \mathfrak{R}, \alpha > 0$ such that

$$l \equiv \lim_{n\to\infty} \left(\sqrt{2n^2 + 4n + 1} - \alpha n - \beta\right) = 2\sqrt{2}.$$

A: $\alpha = \sqrt{2}, \beta = -\sqrt{2}$.

9. Study the convergence of the following series, using simple convergence criterions:

a) $\dfrac{1000}{1!} + \dfrac{1000^2}{2!} + \ldots + \dfrac{1000^n}{n!} + \ldots$ \quad A: convergent

b) $\displaystyle\sum_{n=1}^{\infty} \dfrac{1}{n^2}$ \quad A: convergent

c) $\displaystyle\sum_{n=1}^{\infty} \dfrac{(n!)^2}{(2n)!}$ \quad A: convergent

d) $\sum_{n=1}^{\infty} \frac{n!}{n^n} a^n$,   A: $\begin{cases} \text{converges for } a < e \\ \text{diverges for } a \geq e \end{cases}$

e) $\sum_{n=1}^{\infty} \frac{2^n}{n+a^n}$, $a > 0$ A: $\begin{cases} \text{converges for } a < 2 \\ \text{diverges for } a \geq 2 \end{cases}$

f) $\sum_{n=1}^{\infty} \frac{n^{n+1}}{(2n^2+n+1)^{\frac{n+1}{2}}}$ A: convergent

g) $\sum_{n=1}^{\infty} \left| \tan^n \left(a + \frac{b}{n^2} \right) \right|$, $0 < a < \frac{\pi}{2}$

     A: $\begin{cases} \text{converges for } a \in (0, \pi/4) \\ \text{diverges for } a \in [\pi/4, \pi/2) \end{cases}$

h) $\sum_{n=1}^{\infty} \frac{n^2}{\left(2+\frac{1}{n}\right)^n}$ A: convergent

i) $\frac{1000}{1!} + \frac{1000^2}{2!} + \ldots + \frac{1000^n}{n!} + \ldots$

     A: convergent, the sum is $\frac{2}{3}$

j) $\sum_{n=1}^{\infty} \frac{a(a+1)\ldots(a+n-1)}{b(b+1)\ldots(b+n-1)} x^n$, $a, b, x > 0$, $x \neq 1$

     A: $\begin{cases} \text{converges for } x < 1 \\ \text{diverges for } x > 1 \end{cases}$

k) $\sum_{n=1}^{\infty} \left(\sqrt{n+2} - 2\sqrt{n+1} + \sqrt{n} \right)$

     A: convergent, the sum is $1 - \sqrt{2}$

l) $\dfrac{a+1}{a}b + \left(\dfrac{a+2}{a+1}b\right)^2 + \ldots + \left(\dfrac{a+n}{a+n-1}b\right)^n + \ldots$

A: $\begin{cases} \text{abs. convergent for } |b| < 1 \\ \text{does not converge for } |b| \geq 1 \end{cases}$

m) $1 + \dfrac{1}{1\cdot 4} + \dfrac{1}{2!\cdot 10} + \ldots + \dfrac{1^{n-1}}{n!\cdot(3^n+1)} + \ldots$

A: convergent

n) $\dfrac{1}{8} + \dfrac{1}{18} + \dfrac{1}{28} + \dfrac{1}{38} \ldots$

A: divergent

o) $\dfrac{1}{2\ln 2 \ln\ln 2} + \dfrac{1}{3\ln 3 \ln\ln 3} + \dfrac{1}{n\ln n \ln\ln n} + \ldots$

A: divergent

10. Study the convergence of the following series using special criteria of convergence (Raabe-Duhamel (R-D), logarithm (log), Leibniz):

a) $\sum\limits_{n=1}^{\infty} \dfrac{1\cdot 3\cdot 5\cdot\ldots\cdot(2n-1)}{2\cdot 4\cdot 6\cdot\ldots\cdot 2n}$. A: R-D $\to$ divergent

b) $\sum\limits_{n=1}^{\infty} \dfrac{a(a+1)\ldots(a+n-1)}{b(b+1)\ldots(b+n-1)}$, $a, b > 0$

A: R-D$\to \begin{cases} \text{conv. for } b-a > 1 \\ \text{div. for } b-a \geq 1 \end{cases}$

c) $\sum\limits_{n=1}^{\infty} \dfrac{\sqrt{n!}}{(2+\sqrt{1})(2+\sqrt{2})\cdot\ldots\cdot(2+\sqrt{n})}$

A: R-D $\to$ convergent

d) $\sum_{n=1}^{\infty} \dfrac{1}{n^{\sqrt{n}}}$ A: log → convergent

e) $\sum_{n=1}^{\infty} (-1)^n \dfrac{x^{2n}}{(2n)!}$ A: Leibniz → convergent

f) $\dfrac{a}{b} + \dfrac{a(a+d)}{b(b+d)} + \dfrac{a(a+d)(a+2d)}{b(b+d)(b+2d)} + \ldots$

A: R-D → $\begin{cases} \text{conv. for } \dfrac{b-a}{d} > 1 \\ \text{div. for } \dfrac{b-a}{d} \geq 1 \end{cases}$

Chapter 3

POWER SERIES, SEQUENCES AND SERIES OF FUNCTIONS

3.1. POWER SERIES

Until now, we dealt with number series: $\sum_{n=0}^{\infty} a_n$, $a_n \in \Re$.

To such a series, one can associate, for any $x \in \Re$, a new series

$$a_0 + a_1 x + a_2 x^2 + \ldots + a_n x^n + \ldots \equiv \sum_{n=0}^{\infty} a_n x^n. \qquad (3.1.1)$$

The series (3.1.1) is a *power series*. In a way, it represents a generalization of polynomials.

One can define pointwise the convergence of series (3.1.1), because, for each $x \in \Re$, the series is a numerical one, and for such series we already know what convergence means.

But the sum of the series varies with x and this is why it represents *a function of x*.

Obviously, any power series converges for $x = 0$. But there are series that may fail to converge anywhere on $\Re$, except for $x = 0$.

Example.

The series $1 + 2^2 x + 3^3 x^2 + \ldots + n^n x^{n-1} + \ldots \equiv \sum_{n=0}^{\infty} n^n x^{n-1}$

converges only at $x = 0$. Indeed, its general term $a_n = n^n x^{n-1}$ would tend to 0 (by theorem 2.8), should the series converge, for example, at a point $x = \varepsilon$. But $n^n x^{n-1}$ is not even bounded, no matter what ε is chosen, so the series converges only at $x = 0$.

Yet, being not frequent, these cases are of no interest in practice. In most of the current applications concerning power series, the main problem is to find the domain of convergence of such a series, domain that might be finite or infinite.

3.1.1. DOMAIN OF CONVERGENCE, RADIUS OF CONVERGENCE

Hence, given a power series, we are interested in knowing the set of those $x \in \Re$ for which the series converges; this set is called **the domain of convergence** of the series. In this respect, we can prove

Theorem 3.1 (ABEL). *Let* $\sum_{n=0}^{\infty} a_n x^n$ *be a power series. If for* $x_0 \neq 0$

$$\left| a_n x_0^n \right| < M, \quad n \in \mathfrak{N}, \qquad (3.1.2)$$

then the series $\sum_{n=0}^{\infty} a_n x^n$ *is absolutely convergent on the interval* $(-|x_0|, +|x_0|)$.

Proof. If the inequality (3.1.2) is satisfied, then we also have

$$|a_n x^n| = |a_n x_0^n| \cdot \left|\frac{x}{x_0}\right|^n < M \left|\frac{x}{x_0}\right|^n, \quad n \in \mathfrak{N}. \quad (3.1.3)$$

This inequality shows that the general term of the series $\sum_{n=0}^{\infty} |a_n x^n|$ of moduli is bounded above by the general term of a geometric progression of ratio $\left|\frac{x}{x_0}\right|$. But this ratio is subunitary for $|x| < |x_0|$ and so, by the first part of the comparison theorem, the series of moduli is convergent. In conclusion, the power series is absolutely convergent for $|x| < |x_0|$ and thus, the more so, it is convergent. ◻

Remark. If we can prove that $\sum_{n=0}^{\infty} a_n x_0^n$ converges for a $x_0 \neq 0$, then, obviously, the condition (3.1.2) is satisfied. Indeed, by theorem 2.8, the general term of the series $\sum_{n=0}^{\infty} a_n x_0^n$ tends to 0, hence the sequence $\{a_n x_0^n\}$ is bounded.

Now, let us consider the set of those $x \in \Re$ for which the power series $\sum_{n=0}^{\infty} a_n x^n$ is convergent. According to Abel's theorem, they form a symmetric interval with respect to the origin. Let us note by R the least upper bound of this interval. Obviously, $R > 0$, and

- the power series **converges** for any $x \in (-R, +R)$ (by virtue of the absolute convergence), but

- **does not converge** for $|x| > R$,

because R is the biggest number for which the series converges on the symmetric interval.

At the points $-R, +R$ the series could – or could not – be convergent.

This least upper bound R is called **radius of convergence**.

If $R = 0$, the power series converges only for $x = 0$, and if $R = \infty$, the series converges for any $x \in \Re$.

HOW DO WE FIND THE RADIUS OF CONVERGENCE?

To find intervals of convergence for a given power series $\sum_{n=0}^{\infty} a_n x^n$, we can use, for example, the ratio/root tests, applied to the series of moduli:

$$|a_0| + |a_1 x| + |a_2 x^2| + \ldots + |a_n x^n| + \ldots \equiv \sum_{n=0}^{\infty} |a_n x^n|. \qquad (3.1.4)$$

a) By the **ratio test**, one must have

$$\lim_{n\to\infty}\left|\frac{a_{n+1}x^{n+1}}{a_n x^n}\right| = |x|\lim_{n\to\infty}\left|\frac{a_{n+1}}{a_n}\right| < 1, \qquad (3.1.5)$$

hence the series of moduli converges if

$$|x| < \frac{1}{\lim\limits_{n\to\infty}\left|\dfrac{a_{n+1}}{a_n}\right|}. \qquad (3.1.6)$$

b) By the *root test,* one must have

$$\lim_{n\to\infty}\sqrt[n]{|a_n x^n|} = |x|\sqrt[n]{|a_n|} < 1, \qquad (3.1.7)$$

hence the series of moduli converges if

$$|x| < \frac{1}{\lim\limits_{n\to\infty}\sqrt[n]{|a_n|}}. \qquad (3.1.8)$$

In both cases, if inequalities (3.1.6), (3.1.8) respectively are fulfilled, then the power series is absolutely convergent.

From inequality (3.1.8) we can see that the convergence radius takes at least the value $\dfrac{1}{\lim\limits_{n\to\infty}\sqrt[n]{|a_n|}}$. We will prove that the convergence radius takes **exactly** this value.

It is about a theorem discovered by Cauchy, forgotten for a very long time. Hadamard rediscovered it and pointed out its importance. Nowadays, this theorem, extended to much more general forms, is considered fundamental for spectral theories.

Theorem 3.2 (CAUCHY – HADAMARD). *The radius of convergence of the power series* $\sum_{n=0}^{\infty} a_n x^n$ *is the inverse of the limit superior of the sequence* $\left\{ \sqrt[n]{|a_n|} \right\}_{n \in \mathfrak{N}^*}$, *i.e.*,

$$R = \frac{1}{\rho}, \quad \rho = \overline{\lim_{n \to \infty}} \rho_n, \quad \rho_n = \sqrt[n]{|a_n|}. \tag{3.1.9}$$

* **Proof.** Formula (3.1.9) also contains the limit cases $\rho = 0$, $\rho = \infty$. If $\rho = 0$, then $R = \infty$, and if $\rho = \infty$, then $R = 0$. Thus, we must consider three cases:

A) $\boxed{\rho = 0}$. We will show that $R = \infty$, i.e. the series converges for any $x \in \mathfrak{R}$. Indeed, as $\rho_n = \sqrt[n]{|a_n|}$ tends to 0, the general term of sequence $\sqrt[n]{|a_n x^n|} = |x| \sqrt[n]{|a_n|}$ will also tend to 0, which yields

$$\lim_{n \to \infty} \sqrt[n]{|a_n x^n|} = \lim_{n \to \infty} |x| \sqrt[n]{|a_n|} = 0 < 1; \tag{3.1.10}$$

thus, by the root test (Cauchy), the series $\sum_{n=0}^{\infty} \sqrt[n]{|a_n x^n|}$ is convergent. Consequently, the series $\sum_{n=0}^{\infty} a_n x^n$ is absolutely convergent.

B) $\boxed{\rho = \infty}$. We will show that the series $\sum_{n=0}^{\infty} a_n x^n$ converges only for $x = 0$, which implies $R = 0$.

As $\overline{\lim\limits_{n\to\infty}}\rho_n = \overline{\lim\limits_{n\to\infty}}\sqrt[n]{|a_n|} = \infty$, one can find a subsequence of $\left\{\sqrt[n]{|a_n|}\right\}_{n\in\mathfrak{N}^*}$ which tends to ∞. Let this subsequence be $\left\{\sqrt[n_i]{|a_{n_i}|}\right\}_{n\in\mathfrak{N}}$. Now, let us choose $x \in \mathfrak{R}$, $x \neq 0$. We also have $\lim\limits_{n\to\infty}|x|\sqrt[n_i]{|a_{n_i}|} = \infty$. Hence, there exists an index $N_{|x|}$ such that $\sqrt[n_i]{|a_{n_i}x^{n_i}|} > 1$, $\forall n_i > N_{|x|}$. This means that the series $\sum\limits_{n=0}^{\infty} a_n x^n$ is not even convergent, because its general term does not tend to 0.

C) $\boxed{\rho \in \mathfrak{R}_+, \ \rho < \infty}$ (ρ is a finite positive number). We have already shown that the series converges for $|x| < \dfrac{1}{\rho}$. Now, let us take $|x| > \dfrac{1}{\rho}$. One can find $\varepsilon > 0$ such that $|x| > \dfrac{1}{\rho - \varepsilon}$. According to the property of limit superior, we can find a rank N_ε such that $\sqrt[n]{|a_n|} > \rho - \varepsilon$, $\forall n > N_\varepsilon$ and so

$$\sqrt[n]{|a_n x^n|} = |x|\sqrt[n]{|a_n|} > |x|(\rho - \varepsilon) > 1, \ \forall n > N_\varepsilon, \qquad (3.1.11)$$

wherefrom it results that for such x the power series cannot be convergent.

It means that the convergence radius of the series is exactly $R = \dfrac{1}{\rho}$. ∎

Remark. Relating to numerical sequences, one can prove the following result:

Lemma 3.1. *Let $\{a_n\}_{n\in\mathfrak{N}}$ be a number sequence. If the limit $l_1 \equiv \lim\limits_{n\to\infty}\left|\dfrac{a_{n+1}}{a_n}\right|$ exists and is finite, then $l_2 \equiv \lim\limits_{n\to\infty}\sqrt[n]{|a_n|}$ also exists and $l_1 = l_2$.*

From here, it results that the radius of convergence can be often determined by using the ratio test.

Examples

1. Consider the numerical series of e:

$$1 + \frac{1}{1!} + \frac{1}{2!} + \ldots + \frac{1}{n!} + \ldots .$$

To this series one can associate the power series

$$1 + \frac{x}{1!} + \frac{x^2}{2!} + \ldots + \frac{x^n}{n!} + \ldots . \qquad (3.1.12)$$

Let us find its radius of convergence.

Solution. We have $\rho = \overline{\lim\limits_{n\to\infty}}\sqrt[n]{\dfrac{1}{n!}} = 0$, so $R = \infty$.

The power series (3.1.12) converges for all $x \in \mathfrak{R}$; in fact, its sum is e^x.

2. Let us associate to the harmonic series

$$1 + \frac{1}{2} + \frac{1}{3} \ldots + \frac{1}{n} + \ldots$$ the power series

$$x + \frac{x^2}{2} + \frac{x^3}{3} + \ldots + \frac{x^n}{n} + \ldots . \qquad (3.1.13)$$

Solution. Here, we can apply the ratio test. We compute

$$\lim_{n \to \infty} \left| \frac{a_{n+1}}{a_n} \right| = \lim_{n \to \infty} \frac{\frac{1}{n+1}}{\frac{1}{n}} = 1.$$

Thus, $\rho = 1$, hence $R = 1$. For now, we proved that the series converges on the interval $(-1, +1)$. It remains to see what happens at the ends of the interval.

i) at $x = 1$, the power series becomes $1 + \frac{1}{2} + \frac{1}{3} \ldots + \frac{1}{n} + \ldots$, We thus found the harmonic series, which is ***divergent.***

ii) at $x = -1$, the power series becomes

$$1 - \frac{1}{2} + \frac{1}{3} \ldots + (-1)^{n+1} \frac{1}{n} + \ldots .$$

This is the alternate harmonic series, which is ***convergent.***

Therefore, the domain of convergence of the power series (3.1.13) is the interval $[-1, +1)$.

3. Take the power series

$$1 + x + x^2 + \ldots + x^n + \ldots, \qquad (3.1.14)$$

which can be also considered as *a geometric progression* of ratio x. Hence, the series diverges for $|x| \geq 1$ and converges for $|x| < 1$, its sum being $\dfrac{1}{1-x}$.

The radius of convergence of the series is $R = 1$, and the domain of convergence is the open interval $(-1, +1)$.

3.1.2. OPERATIONS WITH POWER SERIES

Let $\sum_{n=0}^{\infty} a_n x^n$, $\sum_{n=0}^{\infty} b_n x^n$ be two power series.

- *The sum/difference* of the two power series is the series obtained by adding/subtracting the terms of the same rank, i.e.,

$$\sum_{n=0}^{\infty} a_n x^n \pm \sum_{n=0}^{\infty} b_n x^n = \sum_{n=0}^{\infty} (a_n \pm b_n) x^n . \qquad (3.1.15)$$

- A power series is *multiplied by a constant*, multiplying each coefficient by that constant, i.e.,

$$c \sum_{n=0}^{\infty} a_n x^n = \sum_{n=0}^{\infty} c a_n x^n . \qquad (3.1.16)$$

- If $\sum_{n=0}^{\infty} a_n x^n = 0$, $\forall x \in \Re$, then $a_n = 0$, $\forall n \in \mathfrak{N}^*$.

Thus,

The coefficients of the identically zero series are all of them null.

The identically null series is a generalization of the identically zero polynomial.

♣ **The product** of two power series is calculated just as the product of two polynomials.

If $f(x) = \sum_{n=0}^{\infty} a_n x^n$ and $g(x) = \sum_{n=0}^{\infty} b_n x^n$, then

$$f(x)g(x) = (a_0 + a_1 x + \ldots + a_n x^n + \ldots) \cdot$$
$$\cdot (b_0 + b_1 x + \ldots + b_n x^n + \ldots) = \quad (3.1.17)$$
$$= c_0 + c_1 x + c_2 x^2 + \ldots + c_n x^n + \ldots,$$

where

$$c_n = a_0 b_n + a_1 b_{n-1} + \ldots + a_n b_0. \quad (3.1.18)$$

If two series are convergent, their product is also convergent, and its radius of convergence is the least of the convergence radii of the two series.

• **The quotient** of two power series is also a power series.

$$\varphi(x) = \frac{f(x)}{g(x)} = \frac{a_0 + a_1 x + a_2 x^2 + \ldots + a_n x^n + \ldots}{b_0 + b_1 x + b_2 x^2 + \ldots + b_n x^n + \ldots} \quad (3.1.19)$$
$$= \gamma_0 + \gamma_1 x + \ldots + \gamma_n x^n + \ldots .$$

We can compute the quotient if $b_0 \neq 0$. The coefficients γ_n can be determined by identification from the equality $f(x) = \varphi(x) g(x)$. We have

$$a_0 + a_1x + a_2x^2 + \ldots + a_nx^n + \ldots =$$
$$= (\gamma_0 + \gamma_1x + \gamma_2x^2 + \ldots + \gamma_nx^n + \ldots) \cdot \quad (3.1.20)$$
$$\cdot (b_0 + b_1x + b_2x^2 + \ldots + b_nx^n + \ldots).$$

By equating the coefficients of the same powers, we find, step by step

$$b_0\gamma_0 = a_0 \quad \Rightarrow \gamma_0 = \frac{a_0}{b_0},$$

$$b_1\gamma_0 + b_0\gamma_1 = a_1 \quad \Rightarrow \gamma_1 = \frac{a_1 - b_1\gamma_0}{b_0} = \frac{a_1b_0 - a_0b_1}{b_0^2}, \quad (3.1.21)$$

$$b_2\gamma_0 + b_1\gamma_1 + b_0\gamma_2 = a_2$$
$$\Rightarrow \gamma_2 = \frac{a_2 - b_2\gamma_0 - b_1\gamma_1}{b_0} = \ldots.$$

Example. Compute the coefficients of the quotient $\frac{1}{1-x}e^x$.

a) We write the equality

$$1 + \frac{x}{1!} + \frac{x^2}{2!} + \ldots + \frac{x^n}{n!} + \ldots = \quad (3.1.22)$$
$$= (1-x)(\gamma_0 + \gamma_1x + \gamma_2x^2 + \ldots + \gamma_nx^n + \ldots).$$

By identifying the coefficients, we obtain successively

$$\gamma_0 = 1,$$
$$-\gamma_0 + \gamma_1 = \frac{1}{1!} \quad \Rightarrow \gamma_1 = 2, \quad (3.1.23)$$
$$\gamma_2 - \gamma_1 = \frac{1}{2!} \quad \Rightarrow \gamma_2 = \frac{5}{2}, \ldots.$$

b) We can also proceed otherwise, by considering $\dfrac{1}{1-x}$ as the sum of a geometric progression (valid only on $(-1,1)$); thus, we directly effectuate the product

$$\gamma_0 + \gamma_1 x + \gamma_2 x^2 + \ldots + \gamma_n x^n + \ldots =$$
$$= \left(1 + x + x^2 + \ldots + x^n + \ldots\right) \cdot$$
$$\cdot \left(1 + \frac{x}{1!} + \frac{x^2}{2!} + \ldots + \frac{x^n}{n!} + \ldots\right). \quad (3.1.24)$$

We get

$$\gamma_0 = 1,$$
$$\gamma_1 = 1 + \frac{1}{1!} = 2, \quad (3.1.25)$$
$$\gamma_2 = \frac{1}{2!} + \frac{1}{1!} + 1 = \frac{5}{2}, \ldots.$$

Let us note that γ_k coincide by both methods; this should be expected.

3.2. REAL FUNCTIONS, OF ONE REAL VARIABLE

We consider already known the following notions:
- real functions, depending on a real variable;
- continuity, derivability of a function;
- graph of a function;
- derivatives of elementary functions.

Also, it is considered known that, if $f : I = [a,b] \subset \Re \to \Re$ is continuous on I, then

- f is bounded and attains its bounds on I;
- f is uniformly continuous on I.

Let us briefly remind several properties and formulas of calculus for derivatives of functions.

Definition. We say that $f : I \subseteq \Re \to \Re$ is *differentiable at* $x_0 \in I$ if

$$l \equiv \lim_{x \to x_0} \frac{f(x) - f(x_0)}{x - x_0}$$

exists and it is finite; in this case, we write

$$l \equiv f'(x_0).$$

Example. Let $f(x) = x^2$. Obviously, $f : \Re \to \Re$. Let $x_0 \in \Re$ arbitrary. We have

$$\lim_{x \to x_0} \frac{f(x) - f(x_0)}{x - x_0} = \lim_{x \to x_0} \frac{x^2 - x_0^2}{x - x_0} = \lim_{x \to x_0} (x + x_0) = 2x_0.$$

This method of calculus of the derivative has a quite wide degree of generality.

PROPERTIES OF DERIVATIVES

Let $f, g : I \subseteq \Re \to \Re$ be differentiable at $x \in I$. Then

- **the sum** $f + g$ is differentiable at x and

$$\boxed{(f + g)'(x) = f'(x) + g'(x)};$$

- **the product** $f \cdot g$ is differentiable at x and

$$(f \cdot g)'(x) = f'(x)g(x) + g'(x)f(x);$$ → **LEIBNIZ FORMULA**

- if, moreover, $g(x) \neq 0$, then **the ratio** $\dfrac{f}{g}$ is differentiable at x and

$$\left(\dfrac{f}{g}\right)'(x) = \dfrac{f'(x)g(x) - f(x)g'(x)}{g^2(x)};$$

- if $f = f(u)$, $u = u(x)$, are both differentiable with respect to their arguments, the **composite function** $(f \circ u)(x) \equiv f(u(x))$ is differentiable at x and

$$(f \circ g)'(x) = f'(u(x))u'(x).$$

Table 3.1. Derivatives of elementary functions

Nr.	$f(x)$	$f'(x)$	Nr.	$f(x)$	$f'(x)$
1.	x^n	nx^{n-1}	7.	$\sin x$	$\cos x$
2.	e^x	e^x	8.	$\cos x$	$-\sin x$
3.	e^{ax}	ae^x	9.	$\tan x$	$\dfrac{1}{\cos^2 x}$
4.	$\dfrac{1}{x}$	$-\dfrac{1}{x^2}$	10.	$\arctan x$	$\dfrac{1}{1+x^2}$
5.	$\sqrt{x}$	$\dfrac{1}{2\sqrt{x}}$	11.	x^α	$\alpha x^{\alpha-1}$
6.	$\ln|x|$	$\dfrac{1}{x}$	12.	$a^x, a > 0$	$\ln a \cdot a^x$

Table 3.1 contains the most commonly used derivatives of elementary functions.

Consider again the function $f: I \subseteq \Re \to \Re$. We will simplify the presentation and the notations with the following agreements:

- A continuous on *I* function will be called a *function of class* $C^0(I)$, or, otherwise written, $f \in C^0(I)$.

- We say that *f is of class* $C^k(I)$ and we write $f \in C^k(I)$ if *f* is continuous and *k* times differentiable on *I*, and its derivatives $f^{(j)}$, $j = \overline{1,k}$ are continuous on *I*.

- A function of class $C^1(I)$ is also called **smooth on** *I*.

- A function is called *piecewise continuous* if it is continuous on its entire domain of definition, except for a finite number of points, which are points of discontinuity of first kind (removable or jump discontinuities).

- A function is called *piecewise smooth* if it is smooth on its domain of definition, except for a finite number of discontinuity points of first kind.

We remind the reader of several mathematical facts, which are useful for the study of series expansions.

Theorem 3.3. *If $f : I \to \Re$ is differentiable on the open interval I, then its derivative vanishes at a maximum or minimum point.*

Proof. Let $x_0 \in I$ be a maximum point of f. Then, for $h > 0$ we have $f(x_0 + h) < f(x_0)$, so $\dfrac{f(x_0 + h) - f(x_0)}{h} < 0$. But $-h < 0$, hence $\dfrac{f(x_0 - h) - f(x_0)}{-h} > 0$. Taking the limit for $h \to 0$ in the last two inequalities, we obtain simultaneously $f'(x_0) \leq 0$, $f'(x_0) \geq 0$, which implies $f'(x_0) = 0$. ◻

Theorem 3.4 (ROLLE). *If*

i) $f \in C^0([a, b]) \cap C^1((a, b))$ *and*

ii) $f(a) = f(b)$,

then there exists at least one point $c \in (a, b)$ at which the derivative of f vanishes, i.e.,

$$f'(c) = 0. \qquad (3.2.1)$$

* **Proof.** As f is continuous on $[a, b]$ it is bounded and attains its bounds M, m, in $[a, b]$. The points at which these bounds are attained cannot be both extremities a and b at a time. Indeed, in this case we would have $M = m = f(a) = f(b)$, hence the function would be constant on $[a,b]$ and its derivative would be identically 0, i.e., $f'(c) = 0$, $\forall x \in (a,b)$. Except this trivial case, one of the two bounds will be attained at a point $c \in (a,b)$. By theorem 3.3, at this point $f'(c) = 0$. ∎

Theorem 3.5 (CAUCHY). *Consider two functions* $f, \varphi \in C^0([a,b]) \cap C^1((a,b))$ *and suppose that* $\varphi'(x) \neq 0$, $x \in (a,b)$. *Then there exists a point* $c \in (a,b)$ *such that*

$$\boxed{\frac{f(b)-f(a)}{\varphi(b)-\varphi(a)} = \frac{f'(c)}{\varphi'(c)}}. \Rightarrow \text{CAUCHY'S FORMULA} \quad (3.2.2)$$

* **Proof.** Let us set up the auxiliary function

$$F(x) = f(x) - f(a) - \frac{f(b)-f(a)}{\varphi(b)-\varphi(a)}\left[\varphi(x) - \varphi(a)\right]. \quad (3.2.3)$$

We immediately see that $F(a) = 0$, and that

$$F(b) = $$
$$= f(b) - f(a) - \frac{f(b)-f(a)}{\varphi(b)-\varphi(a)}\left[\varphi(b)-\varphi(a)\right] = 0, \quad (3.2.4)$$

so F vanishes at the ends of the interval. On the other hand, by construction, F is a linear combination of f and φ, hence $F \in C^0([a,b]) \cap C^1((a,b))$ too. It follows that F satisfies the hypotheses of Rolle's theorem. It follows that there exists $c \in (a,b)$ such that $F'(c) = 0$. But

$$F'(x) = f'(x) - \frac{f(b)-f(a)}{\varphi(b)-\varphi(a)}\varphi'(x), \qquad (3.2.5)$$

so

$$0 = F'(c) = f'(c) - \frac{f(b)-f(a)}{\varphi(b)-\varphi(a)}\varphi'(c), \qquad (3.2.6)$$

which yields Cauchy's formula (3.2.2). ◘

3.3. LAGRANGE'S FORMULA

Cauchy's formula has a very important particular case, presented in the following theorem.

Theorem 3.6 (LAGRANGE). *Consider the function* $f \in C^0([a,b]) \cap C^1((a,b))$. *Then there exists* $c \in (a,b)$ *such that*

$$\boxed{\frac{f(b)-f(a)}{b-a} = f'(c).} \Rightarrow \textit{LAGRANGE'S FORMULA} \qquad (3.3.1)$$

Proof. Lagrange's theorem can be considered a particular case of Cauchy's theorem. Indeed, we can apply theorem 3.5 choosing as $\varphi(x)$ the identity function, i.e., $\varphi(x) \equiv x$. The identity function obviously satisfies the hypothesis of theorem 3.5. As $\varphi'(x) = 1$, we get at once formula (3.3.1). ◘

Formula (3.3.1) is also called **Lagrange's formula** or *finite increments formula*.

3.3.1. FORMS OF LAGRANGE'S FORMULA

Lagrange's formula (3.3.1) can be written in various forms, appropriate to its multiple applications:

1. $f(b) - f(a) = f'(c)(b-a)$, $c \in (a,b)$;

2. $f(x+h) - f(x) = hf'(x+\theta h)$, $|\theta| < 1$;

3. If $F(x)$ is a primitive of $f(x)$, i.e., $F'(x) = f(x)$, then, by applying Lagrange's formula, it immediately follows that $F(b) - F(a) = f(c)(b-a)$. But $F(b) - F(a) = \int_a^b f(x)dx$, such that, finally we obtain the *integral form of Lagrange's formula.*

$$\boxed{\int_a^b f(x)dx = f(c)(b-a)}. \qquad (3.3.2)$$

3.3.2. THE GEOMETRICAL INTERPRETATION OF LAGRANGE'S FORMULA

Let $\overset{\frown}{AB}$ be the arc representing the graph of the function $f(x)$, which satisfies the hypotheses of the theorem 3.6 (figure 3.1). From the figure, we observe that the gradient of the chord $\overline{AB}$ is $\dfrac{f(b) - f(a)}{b - a}$, and the gradient of the tangent to the curve $\overset{\frown}{ACB}$ at C is $f'(c)$.

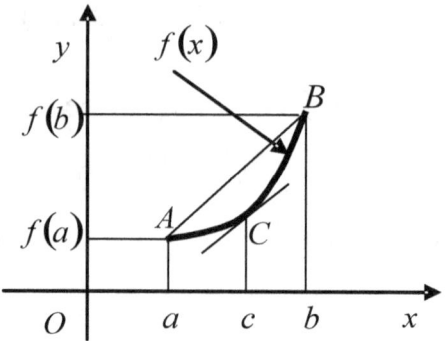

Figure 3.1. The geometrical interpretation of Lagrange's theorem

By virtue of Lagrange's theorem, the two gradients must be equal, hence

The chord $\overline{AB}$ and the tangent to the graph of f at the point C are parallel.

3.3.3. APLICATIONS OF LAGRANGE'S THEOREM

We can prove that

1. The derivative of a function is identically zero if and only if the function is constant.

Indeed, if the function is constant, then it is known that its derivative vanishes identically.

Reciprocally, let f satisfying the hypotheses of Lagrange's theorem and such that $f'(x) = 0$, $x \in I$. For all $x \in I$, we can write Lagrange's formula, which is $f(x) - f(a) = 0 \cdot (b-a)$, so $f(x) = f(a) = \text{const}$.

IMPORTANT CONSEQUENCE:

Two functions whose derivatives coincide differ by a constant.

2. Calculus of approximations

Let $y = f(x)$ be a function expressing the dependence of the physical quantity x, – the independent variable – on the physical quantity y – the dependent variable. Suppose we made some lab experiments which would allow us to get the value of f at a point of interest a. Also suppose that, by our equipment, we cannot fix the variable x precisely at the point a, but only at a point close to it, say a'; this is why we cannot measure exactly $A = f(a)$, but $A' = f(a')$. It is well understood that we can evaluate the difference $|a - a'|$. There arises naturally the following

PROBLEM: *Knowing the difference* $|a - a'| < \varepsilon$, *evaluate* $|A - A'|$.

Solution. We apply Lagrange's formula, provided the function is differentiable. We have

$$|A - A'| = |f(a) - f(a')| =$$
$$= |f'(a' + \theta(a - a'))||a - a'| < M|a - a'|, \quad (3.3.3)$$
$$\theta \in (0,1),$$

where $M = \sup\limits_{|x-a'|<\varepsilon} |f'(x)|$. If M is known, it follows that the difference

$$|A - A'| \leq M\varepsilon \quad (3.3.4)$$

can also be evaluated.

Example. Let $f(x) = \ln(1+x)$. Let us evaluate the positive difference

$$\ln 101 - \ln 100.$$

Solution. The derivative of f is

$$f'(x) = \frac{1}{1+x}.$$

We have

$$\ln 101 - \ln 100 = \ln \frac{101}{100} = \ln\left(1 + \frac{1}{100}\right).$$

So $a = 0$, $a' = \frac{1}{100}$. Further,

$$\left|\ln\left(1 + \frac{1}{100}\right) - \ln 1\right| \le \frac{1}{1 + \frac{\theta}{100}} \cdot \left|\left(1 + \frac{1}{100}\right) - 1\right|, \quad \theta \in (0,1),$$

such that

$$\ln 101 - \ln 100 < \frac{1}{100 + \theta} < \frac{1}{100} = 0.01.$$

Indeed,

$$\left.\begin{array}{l}\ln 100 \cong 4,60517 \\ \ln 101 \cong 4.61512\end{array}\right| \Rightarrow \ln 101 - \ln 100 \cong 0.00995.$$

3.4. TAYLOR'S FORMULA

Consider a function $f \in C^2([a,b]) \cap C^3((a,b))$ and let C be the constant defined by the formula

$$f(b) = f(a) + \frac{(b-a)}{1!} f'(a) +$$
$$+ \frac{(b-a)^2}{2!} f''(a) + (b-a)^p C, \quad (3.4.1)$$

where $p \leq 3$ is a natural number.

Also consider the auxiliary function

$$F(x) = f(x) + \frac{(b-x)}{1!} f'(x) +$$
$$+ \frac{(b-x)^2}{2!} f''(x) + (b-x)^p C. \quad (3.4.2)$$

We observe that

i) $F \in C^0([a,b]) \cap C^1((a,b))$;

ii) $\begin{cases} F(b) = f(b) \\ F(a) = f(b) \end{cases} \Rightarrow F(a) = F(b).$

It follows that F satisfies the hypotheses of Rolle's theorem's, hence there exists $\lambda \in (a,b)$ such that $F'(\lambda) = 0$.

We differentiate formula (3.4.2)

$$F'(x) = f'(x) + \frac{(b-x)}{1!} f''(x) + \frac{(b-x)^2}{2!} f'''(x) -$$
$$- f'(x) - \frac{(b-x)}{1!} f''(x) - p(b-x)^{p-1} C, \quad (3.4.3)$$

therefore

$$F'(x) = \frac{(b-x)^2}{2!} f'''(x) - p(b-x)^{p-1} C. \quad (3.4.4)$$

But $F'(\lambda) = 0$ involves

$$C = \frac{(b-\lambda)^2}{2! \, p (b-x)^{p-1}} f'''(\lambda), \qquad (3.4.5)$$

or

$$C = \frac{(b-\lambda)^{3-p}}{2! \, p} f'''(\lambda). \qquad (3.4.6)$$

Introducing this expression of C in (3.4.2), we find

$$\boxed{\begin{aligned} f(b) = f(a) + \frac{(b-a)}{1!} f'(a) + \\ + \frac{(b-a)^2}{2!} f''(a) + \frac{(b-a)^p (b-\lambda)^{3-p}}{2! \, p} f'''(\lambda), \end{aligned}} \qquad (3.4.7)$$

which represents **TAYLOR'S FORMULA** for $n = 3$.

For any natural n, we can similarly prove

Theorem 3.7 (TAYLOR). *Let f be a function belonging to $C^{n-1}([a,b]) \cap C^n((a,b))$. Then the following formula is valid*

$$\boxed{\begin{aligned} f(b) = f(a) + \frac{(b-a)}{1!} f'(a) + \frac{(b-a)^2}{2!} f''(a) + \ldots \\ + \frac{(b-a)^{n-1}}{(n-1)!} f^{(n-1)}(a) + \\ + \frac{(b-a)^p (b-\lambda)^{n-p}}{p(n-1)!} f^{(n)}(\lambda), \; \lambda \in (a,b), \; p \leq n, \end{aligned}} \qquad (3.4.8)$$

*also known as **TAYLOR'S FORMULA** for arbitrary natural n.*

From the above considerations, it follows that Rolle's formula is the foundation of Lagrange's, Cauchy's and Taylor's formulas, which are currently used in applications. Intuitively, we can represent the relative positions of theorems 3.4, 3.6 and 3.7 in the form

ROLLE'S THEOREM<< LAGRANGE'S THEOREM << TAYLOR'S THEOREM,

where the "order relation << " has the sense of a simpler corresponding formula.

Particular case: for $n = 1$, $p = 1$, Taylor's formula (3.4.8) becomes

$$\boxed{f(b) = f(a) + \frac{b-a}{1 \cdot 1!} f'(\lambda), \quad \lambda \in (a,b)}, \qquad (3.4.9)$$

i.e., Lagrange's formula.

3.4.1. THE REMAINDER OF TAYLOR'S FORMULA

The expression $R_{np} = \dfrac{(b-a)^p (b-\lambda)^{n-p}}{p(n-1)!} f^{(n)}(\lambda)$

from (3.4.8) is called *the remainder of Taylor's formula*.

Forms of the remainder

1. By taking $\lambda = a + \theta(b-a)$, $0 < \theta < 1$, we obtain

$$b - \lambda = b - a - \theta(b-a) = (b-a)(1-\theta), \qquad (3.4.10)$$

wherefrom

$$(b-a)^p (b-\lambda)^{n-p} = (b-a)^{p+n-p}(1-\theta)^p =$$
$$= (b-a)^n (1-\theta)^p, \qquad (3.4.11)$$

hence

$$\boxed{R_{np} = \frac{(b-a)^n (1-\theta)^p}{p(n-1)!} f^{(n)}(a+(b-a)\theta)}, \qquad (3.4.12)$$

which is **SCHLÖMLICH'S REMAINDER.**

2. By taking $p - n$ in formula (3.4.12), we obtain **LAGRANGE'S REMAINDER:**

$$\boxed{R_n = \frac{(b-a)^n}{n!} f^{(n)}(\lambda)}. \qquad (3.4.13)$$

3. By taking $p = 1$ in formula (3.4.12), we obtain **CAUCHY'S REMAINDER:**

$$\boxed{R_{n1} = \frac{(b-a)(b-\lambda)^{n-1}}{(n-1)!} f^{(n)}(\lambda)}. \qquad (3.4.14)$$

3.4.2. VARIANTS OF TAYLOR'S FORMULA

To write them, we use Lagrange's remainder.

1. In the general Taylor's formula, we replace the remainder R_{np} with Lagrange's remainder, thus getting

$$f(b) = f(a) + \frac{(b-a)}{1!} f'(a) + \frac{(b-a)^2}{2!} f''(a) +$$
$$+ \ldots + \frac{(b-a)^{n-1}}{(n-1)!} f^{(n-1)}(a) + \frac{(b-a)^n}{n!} f^{(n)}(\lambda), \quad (3.4.15)$$
$$\lambda \in (a, b).$$

2. Putting $a = x$, $b = x + h$ in (3.4.8), which yields $b - a = x + h - x = h$, we infer

$$f(x+h) = f(x) + \frac{h}{1!} f'(x) + \frac{h^2}{2!} f''(x) + \ldots$$
$$+ \frac{h^{n-1}}{(n-1)!} f^{(n-1)}(x) + \frac{h^n}{n!} f^{(n)}(x + \theta h), \quad 0 < \theta < 1; \quad (3.4.16)$$

this form is frequently used in the ***theory of aproximations***.

3. Putting $b = x$ in (3.4.8), we obtain

$$f(x) = f(a) + \frac{(x-a)}{1!} f'(a) +$$
$$+ \frac{(x-a)^2}{2!} f''(a) + \ldots + \frac{(x-a)^{n-1}}{(n-1)!} f^{(n-1)}(a) + \quad (3.4.17)$$
$$+ \frac{(x-a)^n}{n!} f^{(n)}(a + (x-a)\theta), \quad 0 < \theta < 1.$$

3.4.3. TAYLOR POLYNOMIAL

In this variant of Taylor's formula, one can distinguish a polynomial $P_{n-1}(x)$ of a single variable x, more precisely

$$P_{n-1}(x) \equiv f(a) + \frac{(x-a)}{1!} f'(a) + \frac{(x-a)^2}{2!} f''(a) +$$
$$+ \ldots + \frac{(x-a)^{n-1}}{(n-1)!} f^{(n-1)}(a). \tag{3.4.18}$$

This is the **TAYLOR POLYNOMIAL**.

From formula (3.4.17) we see that, if the remainder tends to zero, the function can be approximated in a neighbourhood of a by the Taylor polynomial:

$$f(x) \cong P_{n-1}(x). \tag{3.4.19}$$

Let us write the Taylor formula (3.4.17) if the function $f(x)$ is a n degree polynomial, i.e.

$$f(x) \equiv P_n(x) = a_0 + a_1 x + a_2 x^2 + \ldots + a_n x^n. \tag{3.4.20}$$

To write the remainder, we must calculate $P_n^{(n)}(x) = n! a_n \equiv \text{const}, \forall x \in \Re$. The Lagrange remainder becomes in this case

$$R_n = \frac{(x-a)^n}{n!} n! a_n = (x-a)^n a_n. \tag{3.4.21}$$

So, the Taylor formula for a polynomial reads

$$P_n(x) \equiv P_n(a) + \frac{(x-a)}{1!} P_n'(a) + \frac{(x-a)^2}{2!} P_n''(a) + \ldots +$$
$$+ \frac{(x-a)^m}{m!} P_n^m(a) + \ldots + \frac{(x-a)^{n-1}}{(n-1)!} P_n^{(n-1)}(a) + \tag{3.4.22}$$
$$+ (x-a)^n a_n.$$

This form of the polynomial is helpful in the study of multiple roots.

Multiple roots

Suppose that a is a multiple root of order m for $P_n(x)$. Then

$$P_n(x) = (x-a)^m Q_{n-m}(x), \quad Q_{n-m}(a) \neq 0, \quad (3.4.23)$$

where Q_{n-m} is a $n-m$ degree polynomial, which is not divisible by $(x-a)$.

As a has the order m of multiplicity, the formulas (3.4.22) and (3.4.23) must coincide, hence one must have

$$\begin{aligned} & P_n(a) = 0, \quad P'_n(a) = 0, \quad P''_n(a) = 0, \\ & P_n^{(m-1)}(a) = 0, \quad P_n^{(m)}(a) \neq 0. \end{aligned} \quad (3.4.24)$$

It follows that:

$P_n(x)$ *allows a as a multiple root of order m if and only if*

$$P_n^{(k)}(a) = 0, \quad k = \overline{0, m-1}, \text{ but } P_n^{(m)}(a) \neq 0. \quad (3.4.25)$$

Particularly, if α is a double root of the polynomial $ax^2 + bx + c$, then

$$\begin{cases} \alpha^2 + b\alpha + c = 0, \\ 2a\alpha + b = 0, \end{cases} \quad (3.4.26)$$

whence

$$a\left(-\frac{b}{2a}\right)^2 + b\left(-\frac{b}{2a}\right) + c = 0. \quad (3.4.27)$$

Finally, in the case of a double root, the coefficients of the quadratic polynomial must satisfy the well-known condition

$$b^2 - ac = 0. \qquad (3.4.28)$$

3.5. POWER SERIES EXPANSIONS

The Taylor formula, actually, serves at approximating of functions by Taylor polynomials, if the remainder R_{np} can be made small enough. If $\lim_{n \to \infty} R_{np} = 0$, then we can even obtain **the exact value of** f around a as the sum of **the Taylor series**

$$\boxed{\begin{aligned} f(x) = f(a) &+ \frac{(x-a)}{1!} f'(a) + \frac{(x-a)^2}{2!} f''(a) + \\ &+ \ldots + \frac{(x-a)^n}{n!} f^{(n)}(a) + \ldots . \end{aligned}} \qquad (3.5.1)$$

$$\Downarrow$$

TAYLOR SERIES EXPANSION

which is a power series.

For $a = 0$, the Taylor series becomes a Mac Laurin series:

$$\boxed{\begin{aligned} f(x) = f(0) &+ \frac{x}{1!} f'(0) + \frac{x^2}{2!} f''(0) + \ldots + \\ &+ \frac{x^n}{n!} f^{(n)}(0) + \ldots . \end{aligned}} \qquad (3.5.2)$$

$$\Downarrow$$

MAC LAURIN SERIES

Theorem 3.8. Let $f \in C^{\infty}(I)$, $I = [a, b] \subset \mathfrak{R}$, and

$$\sup_{x \in I} \left| f^{(n)}(x) \right| \leq M, \ \forall n \in \mathfrak{N}^*. \qquad (3.5.3)$$

Then f allows a Taylor series expansion around every $x \in I$.

Proof. Consider the Lagrange remainder

$$R_n = \frac{(x-a)^n}{n!} f^{(n)}(\lambda), \ x \in I, \ \lambda \in (a, x). \qquad (3.5.4)$$

From the hypothesis, it results that

$$|R_n| < M \frac{|x-a|^n}{n!}, \ x \in I, \qquad (3.5.5)$$

hence $\lim_{n \to \infty} R_n = 0$. ∎

3.6. MAC LAURIN SERIES FOR ELEMENTARY FUNCTIONS

1. EXPONENTIAL FUNCTION: $f(x) = e^x$. For any $n \in \mathfrak{N}$, $f^{(n)}(x) = e^x$, so $f^{(n)}(0) = 1$, $\forall n \in \mathfrak{N}$. It follows that

$$\boxed{e^x = 1 + \frac{x}{1!} + \frac{x^2}{2!} + \frac{x^3}{3!} + \ldots + \frac{x^n}{n!} + \ldots} \qquad (3.6.1)$$

⇓

MAC LAURIN SERIES OF THE EXPONENTIAL FUNCTION

By taking $x = 1$, we obtain again the series of e.

2. TRIGONOMETRIC FUNCTIONS:

$f(x) = \sin x$, $g(x) = \cos x$.

We have

$$\begin{cases} f(x) = \sin x, \\ f'(x) = \cos x, \\ f''(x) = -\sin x, \\ f'''(x) = -\cos x, \\ f^{IV}(x) = \sin x = f(x), \end{cases} \quad \begin{cases} g(x) = \cos x, \\ g'(x) = -\sin x, \\ g''(x) = -\cos x, \\ g'''(x) = \sin x, \\ g^{IV}(x) = \cos x = g(x), \end{cases} \quad (3.6.2)$$

hence

$$\begin{aligned} f^{(4k)}(0) &= 0, & g^{(4k)}(0) &= 1, \\ f^{(4k+1)}(0) &= 1, & g^{(4k+1)}(0) &= 0, \\ f^{(4k+2)}(0) &= 0, & g^{(4k+2)}(0) &= -1, \\ f^{(4k+3)}(0) &= -1, & g^{(4k+3)}(0) &= 0. \end{aligned} \quad (3.6.3)$$

It follows that the Mac Laurin series of $\sin x$, $\cos x$ are

$$\boxed{\begin{aligned} \sin x &= \frac{x}{1!} - \frac{x^3}{3!} + \frac{x^5}{5!} - \frac{x^7}{7!} \cdots, \\ \cos x &= 1 - \frac{x^2}{2!} + \frac{x^4}{4!} - \frac{x^6}{6!} + \cdots \end{aligned}} \quad (3.6.4)$$

⇓

MAC LAURIN SERIES FOR $\sin x$, $\cos x$

or, using summation,

$$\boxed{\begin{aligned} \sin x &= \sum_{n=1}^{\infty} (-1)^{(n-1)} \frac{x^{2n-1}}{(2n-1)!}, \\ \cos x &= \sum_{n=0}^{\infty} (-1)^{(n)} \frac{x^{2n}}{(2n)!}. \end{aligned}} \quad (3.6.5)$$

It is easily seen that the sinus expansion contains only odd, while that of the cosinus – only even powers of x. Consequently, according to their expansions too, it follows that:

- $\sin x$ is an ***odd function***,
- $\cos x$ is an ***even function***.

a) APPROXIMATING TRIGONOMETRIC FUNCTIONS

For small values of x, we can aproximate sin and cos by the first terms of their Mac Laurin expansion respectively:

$$\sin x \cong x, \quad \cos x \cong 1 - \frac{x^2}{2}. \tag{3.6.6}$$

For $x > 0$, the series of $\sin x$ is an alternating series, therefore its sum is always found between the sum of an even and the sum of an odd number of terms:

$$x - \frac{x^3}{3!} < \sin x < x. \tag{3.6.7}$$

ATENTION!!

In Taylor's series of trigonometric functions, the argument x is taken in RADIANS, not in DEGREES!

Example. Aproximate $\sin 1^0$.

Solution. We write the argument in radians:

$$x = 1^0 = \frac{\pi}{180} = 0.017453292\ldots .$$

We calculate the successive powers of the argument

$$\frac{x^2}{2!} = 0.00152309...; \quad \frac{x^3}{3!} = 0.000000886...;$$

$$\frac{x^4}{4!} = 0.00000004...; \quad \frac{x^5}{5!} = 0.000000000... .$$

$$\boxed{\begin{aligned} \sin 1^0 &= 0.017452406..., \\ \cos 1^0 &= 0.999847695... . \end{aligned}}$$

If we wish to compute with a given precision, we must consider Taylor's formula and its remainder.

b) EULER'S FORMULAS

Let us expand the function e^{ix} in a Mac Laurin series (see section 3.7.7, dedicated to power series in complex):

$$e^{ix} = 1 + \frac{ix}{1!} + \frac{(ix)^2}{2!} + \frac{(ix)^3}{3!} + \frac{(ix)^4}{4!} ... + \frac{(ix)^n}{n!} + ... \quad (3.6.8)$$

Keeping in mind operations with i, we will have

$$i^2 = -1, \quad i^3 = -i, \quad i^4 = +1, \quad (3.6.9)$$

so

$$i^{4k+1} = i, \quad i^{4k+2} = -1, \quad i^{4k+3} = -i, \quad i^{4k} = +1. \quad (3.6.10)$$

The above mentioned series reads

$$e^{ix} = 1 + \frac{ix}{1!} - \frac{x^2}{2!} + \frac{ix^3}{3!} + \frac{x^4}{4!} + \frac{ix^5}{5!} - \frac{x^6}{6!} + ...; \quad (3.6.11)$$

separating the real and the imaginary part, we find

$$e^{ix} = \underbrace{1 - \frac{x^2}{2!} + \frac{x^4}{4!} - \frac{x^6}{6!} + \ldots}_{\cos x} +$$

$$+ i\left(\underbrace{\frac{x}{1!} - \frac{x^3}{3!} + \frac{x^4}{4!} - \frac{x^5}{5!} + \ldots}_{\sin x}\right), \qquad (3.6.12)$$

hence

$$e^{ix} = \cos x + i \sin x. \qquad (3.6.13)$$

Taking the conjugate, it results

$$e^{-ix} = \cos x - i \sin x. \qquad (3.6.14)$$

From the last two formulas, we easily get

$$\boxed{\begin{aligned} \cos x &= \frac{e^{ix} + e^{-ix}}{2}, \\ \sin x &= \frac{e^{ix} - e^{-ix}}{2i}. \end{aligned}} \rightarrow \textbf{EULER'S FORMULAS} \qquad (3.6.15)$$

3. HYPERBOLIC FUNCTIONS

To introduce them, we consider the Mac Laurin series of e^x, e^{-x}, that we add/subtract respectively:

$$\left.\begin{aligned} e^{x} &= 1 + \frac{x}{1!} + \ldots + \frac{x^{2n-1}}{(2n-1)!} + \frac{x^{2n}}{(2n)!} + \ldots \\ e^{-x} &= 1 - \frac{x}{1!} + \ldots - \frac{x^{2n-1}}{(2n-1)!} + \frac{x^{2n}}{(2n)!} - \ldots \end{aligned}\right| +/-$$

$$e^x + e^{-x} = 2\left(1 + \frac{x^2}{2!} + \frac{x^4}{4!} + \ldots + \frac{x^{2n}}{(2n)!} + \ldots\right) \qquad (3.6.16)$$

$$e^x - e^{-x} = 2\left(\frac{x}{1!} + \frac{x^3}{3!} + \ldots \frac{x^{2n-1}}{(2n-1)!} + \ldots\right).$$

The operations performed above are legitimate because of the convergence of the exponential series on $\Re$.

The hyperbolic functions are defined as follows:

$$\begin{cases} \cosh x = \dfrac{e^x + e^{-x}}{2} \to \textbf{\textit{hyperbolic cosinus}}\,(\text{ch}\,x), \\ \sinh x = \dfrac{e^x - e^{-x}}{2} \to \textbf{\textit{hyperbolic sinus}}\,(\text{sh}\,x). \end{cases} \qquad (3.6.17)$$

From the expansions (3.6.16), it follows that the series for $\sinh x, \cosh x$ are:

$$\cosh x = 1 + \frac{x^2}{2!} + \frac{x^4}{4!} + \ldots + \frac{x^{2n}}{(2n)!} + \ldots,$$

$$\sinh x = \frac{x}{1!} + \frac{x^3}{3!} + \ldots \frac{x^{2n-1}}{(2n-1)!} + \ldots . \qquad (3.6.18)$$

We observe that:
- $\cosh x > 1$ for $x \neq 0$ and $\cosh x = 1 \Leftrightarrow x = 0$;
- $\sinh x > 0$ for $x > 0$, $\sinh x < 0$ for $x < 0$ and $\sinh x = 0 \Leftrightarrow x = 0$.

a) ANALOGIES WITH TRIGONOMETRIC FUNCTIONS:

- the function cosh is *even*, and
- the function sinh is *odd*,

as in the case of trigonometric functions.

Indeed, we immediately observe that, both from the definition formulas, and from the corresponding series,

$$\cosh(-x) = \cosh x, \quad \sinh(-x) = -\sinh x. \qquad (3.6.19)$$

We can easily prove the following formula

$$\boxed{\cosh^2 x - \sinh^2 x = 1, \, x \in \Re}. \qquad (3.6.20)$$

Indeed,

$$\cosh^2 x - \sinh^2 x = \left(\frac{e^x + e^{-x}}{2}\right)^2 - \left(\frac{e^x - e^{-x}}{2}\right)^2 = \qquad (3.6.21)$$

$$= \frac{1}{4}\left(e^{2x} + 2 + e^{2x} - e^{2x} + 2 - e^{2x}\right) = 1.$$

b) WHY ARE THEY CALLED HYPERBOLIC FUNCTIONS??

If

$$\begin{cases} x = \cosh t, \\ y = \sinh t, \end{cases} \qquad (3.6.22)$$

we see from (3.6.20) that x and y satisfy the condition

$$x^2 - y^2 = 1, \qquad (3.6.23)$$

hence the point (x, y) belongs to an *equilateral hyperbola.*

In the case of trigonometric functions, the point (x, y), where

$$\begin{cases} x = \cos t, \\ y = \sin t, \end{cases} \qquad (3.6.24)$$

is on the unit circle

$$x^2 + y^2 = 1. \qquad (3.6.25)$$

c) USUAL FORMULAS:

$$\begin{cases} \cosh(x \pm y) = \cosh x \cosh y \pm \sinh x \sinh y, \\ \sinh(x \pm y) = \sinh x \cosh y \pm \cosh x \sinh y. \end{cases} \qquad (3.6.26)$$

d) DERIVATIVES OF HYPERBOLIC FUNCTIONS:

$$\frac{d}{dx}(\cosh x) = \frac{1}{2}\frac{d}{dx}(e^x + e^{-x}) = \frac{1}{2}(e^x - e^{-x}) =$$
$$= \sinh x, \qquad (3.6.27)$$
$$\frac{d}{dx}(\sinh x) = \cosh x.$$

e) HYPERBOLIC TANGENT AND COTANGENT
are defined as follows:

$$\tanh x = \frac{\sinh x}{\cosh x}, \quad \coth x = \frac{\cosh x}{\sinh x}. \qquad (3.6.28)$$

f) APLICATION IN MECHANICS:

The catenary

The curve $y = C \cosh ax$ is called *catenary*, because it represents the equilibrium position of a simply supported elastic bar, isotropic and homogeneous.

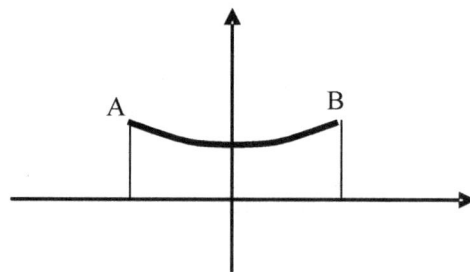

Figure 3.2. The catenary

4. THE GENERALISED BINOMIAL SERIES

Consider the function

$$f(x) = (1+x)^\lambda, \; |x| < 1, \qquad (3.6.29)$$

where $\lambda \in \Re$ is arbitrary. To write the associated Mac Laurin series, we must

1) compute the successive derivatives of the function at $x = 0$ and

2) study the remainder.

1) We have successively

$$f'(x) = \lambda(1+x)^{\lambda-1},$$
$$f''(x) = \lambda(\lambda-1)(1+x)^{\lambda-2}, \quad (3.6.30)$$
$$\dots\dots\dots\dots\dots\dots\dots\dots\dots\dots\dots\dots$$
$$f^{(n)}(x) = \lambda(\lambda-1)\dots(\lambda-n+1)(1+x)^{\lambda-n}, \dots.$$

Taking the values at the origin, it follows that

$$f^{(n)}(0) = \lambda(\lambda-1)\dots(\lambda-n+1). \quad (3.6.31)$$

We thus formally obtain

$$\boxed{\begin{aligned}(1+x)^{\lambda} &= 1 + \lambda\frac{x}{1!} + \lambda(\lambda-1)\frac{x^2}{2!} + \\ &+ \dots + \lambda(\lambda-1)\dots(\lambda-n+1)\frac{x^n}{n!} + \dots.\end{aligned}} \quad (3.6.32)$$

$$\Downarrow$$

THE GENERALISED BINOMIAL SERIES

* 2) To study the convergence, it is suitable to use the Cauchy form of the remainder, i.e.

$$R_{n1} = \frac{(b-a)(b-\mu)^{n-1}}{(n-1)!} f^{(n)}(\mu), \quad (3.6.33)$$

where we take $a = 0$, $b = x$, $\mu = \theta x$, with $0 < \theta < 1$. Thus,

$$R_{n1} = \frac{x^n (1-\theta)^{n-1}}{(n-1)!} f^{(n)}(\theta x), \qquad (3.6.34)$$

or, by replacing the derivative,

$$\begin{aligned} R_{n1} &= \\ &= \frac{x^n (1-\theta)^{n-1}}{(n-1)!} \lambda(\lambda-1)\ldots(\lambda-n+1)(1+\theta x)^{\lambda-n} = \\ &= \underbrace{\frac{\lambda(\lambda-1)\ldots(\lambda-n+1)}{(n-1)!} x^n}_{u_n} (1-\theta)^{n-1} (1+\theta x)^{\lambda-n}. \end{aligned} \qquad (3.6.35)$$

We obtain

$$R_{n1} = u_n (1-\theta)^{n-1} (1+\theta x)^{\lambda-n}, \qquad (3.6.36)$$

and, finally,

$$R_{n1} = u_n \left(\frac{1-\theta}{1+\theta x}\right)^{n-1} (1+\theta x)^{\lambda-1}. \qquad (3.6.37)$$

Let us consider the series $\sum_{n=1}^{\infty} u_n$. Applying the ratio criterion, it follows that this series is absolutely convergent for any $\lambda \in \Re$, if $|x| < 1$. Indeed,

$$\lim_{n\to\infty} \left|\frac{u_{n+1}}{u_n}\right| = \lim_{n\to\infty} \left|\frac{\frac{\lambda(\lambda-1)\ldots(\lambda-n+1)(\lambda-n)}{n!} x^{n+1}}{\frac{\lambda(\lambda-1)\ldots(\lambda-n+1)}{(n-1)!} x^n}\right| = \\ = \lim_{n\to\infty} \left|\frac{\lambda-n}{n}\right| |x| = |x| < 1, \qquad (3.6.38)$$

which involves the absolute convergence of $\sum_{n=1}^{\infty} u_n$.

Then, by theorem 2.8, it results that $\lim_{n\to\infty} u_n = 0$.

Now, let us take the other factors from (3.6.36). The last one is bounded, because of the inequality

$$(1-|x|)^{\lambda-1} \leq (1+\theta x)^{\lambda-1} \leq (1+|x|)^{\lambda-1}, \qquad (3.6.39)$$

hance

- for $\lambda > 1$, $(1+\theta x)^{\lambda-1} \leq (1+|x|)^{\lambda-1} < 2^{\lambda-1}$, and
- for $\lambda < 1$, $(1+\theta x)^{\lambda-1} \leq (1-|x|)^{\lambda-1} < 1$.

The ratio

$$\left|\frac{1-\theta}{1+\theta x}\right| \leq \frac{1-\theta}{1-\theta|x|} < 1 \qquad (3.6.40)$$

is bounded. In conclusion, $\lim_{n\to\infty} R_{n1} = 0$. Therefore, the series (3.6.32) converges for $|x| < 1$, $\forall \lambda \in \mathfrak{R}$.

One can show that the generalised binomial series is absolutely convergent in the following cases:

i) $|x| < 1$, $\forall \lambda \in \mathfrak{R}$ – previously proved;

ii) $|x| \leq 1$, $\lambda > -1$.

WHY GENERALISED BINOMIAL?

If $\lambda = n$, then $f(x) = (1+x)^{\lambda}$ becomes Newton's classical binomial. According to the well-known formula from high school, we have

$$(1+x)^n = 1 + C_n^1 x + C_n^2 x^2 + \ldots + C_n^k x^k + \ldots + x^n. \quad (3.6.41)$$

On the other hand, if we apply formula (3.6.32), the series interrupts at the term of order n, because $\lambda - n = 0$.

We obtain

$$(1+x)^n = 1 + n\frac{x}{1!} + n(n-1)\frac{x^2}{2!} + \ldots +$$
$$+ n(n-1)\ldots(n-k+1)\frac{x^k}{k!} + \ldots + n(n-1)\ldots 2 \cdot 1 \frac{x^n}{n!} =$$
$$= 1 + \frac{n}{1!}x + \frac{n(n-1)}{2!}x^2 + \ldots \quad (3.6.42)$$
$$+ \frac{n(n-1)\ldots(n-k+1)}{k!}x^k + \ldots + x^n,$$

which coincides with Newton's binomial.

Example. Compute $\sqrt[5]{100005}$ with 4 decimal places.

Solution. We have

$$\sqrt[5]{100005} = \sqrt[5]{100000 + 5} =$$
$$= \sqrt[5]{100000\left(1 + \frac{5}{10^5}\right)} = 10\left(1 + \frac{5}{10^5}\right)^{\frac{1}{5}}.$$

So,

$$x = \frac{5}{10^5}, \quad \lambda = \frac{1}{5}.$$

To obtain the desired precision, we estimate the corresponding Lagrange's remainder.

$$R_n = \frac{\left(\frac{5}{10^5}\right)^n}{n!} \frac{1}{5}\left(\frac{1}{5}-1\right)\left(\frac{1}{5}-2\right)\left(\frac{1}{5}-n+1\right)\left(1+\theta\frac{5}{10^5}\right)^{\frac{1}{5}-n},$$

hence

$$|R_n| \le \left(\frac{5}{10^5}\right)^n \left|\frac{1}{5}\left(\frac{1}{5}-1\right)\left(\frac{1}{10}-1\right)\cdots\left(\frac{1}{5(n-1)}-1\right)\frac{1}{n}\right|$$

$$\le \left(\frac{5}{10^5}\right)^n \frac{1}{5n}.$$

To ensure the desired precision, we must take n such that

$$\left(\frac{5}{10^5}\right)^n \frac{1}{5n} < 0.00001.$$

So, $n \ge 1$. By taking $n = 2$, we obtain

$$\sqrt[5]{100005} \cong 10\left(1 + \frac{1}{5}\frac{5}{10^5} + \frac{1}{2!}\left(\frac{5}{10^5}\right)^2 \frac{1}{5}\left(\frac{1}{5}-1\right)\right) =$$

$$= 10.0001 - 4 \cdot 10^{-10}.$$

It means that $n = 1$ ensures a precision of 10^{-4}.

5. THE LOGARITHM SERIES EXPANSION

Consider the function

$$f(x) = \ln(1+x), \quad |x| < 1. \tag{3.6.43}$$

We calculate its successive derivatives:

$$\frac{d}{dx}\ln(1+x) = \frac{1}{1+x} = (1+x)^{-1},$$

$$\frac{d}{dx}(1+x)^{-1} = -(1+x)^{-2},$$

$$\frac{d^2}{dx^2}(1+x)^{-2} = +2.1(1+x)^{-2}, \qquad (3.6.44)$$

$$\cdots\cdots\cdots\cdots\cdots\cdots\cdots\cdots\cdots\cdots\cdots\cdots\cdots$$

$$\frac{d^n}{dx^n}(1+x)^{-n} = +n(n-1)\ldots 2.1(1+x)^{-n-1},\ldots$$

hence

$$\left.\frac{d}{dx}\ln(1+x)\right|_{x=0} = 1,$$

$$\left.\frac{d}{dx}(1+x)^{-1}\right|_{x=0} = -1, \qquad (3.6.45)$$

$$\left.\frac{d^n}{dx^n}(1+x)^{-n}\right|_{x=0} = (-1)^n n!,\ldots .$$

We thus obtain the Mac Laurin series expansion

$$\ln(1+x) = 0 + \frac{x}{1!}.1 - \frac{x^2}{2!}.1 + \frac{x^3}{3!}.2! + \ldots$$
$$+ (-1)^{n+1}\frac{x^n}{n!}.(n-1)! + \ldots, \qquad (3.6.46)$$

which becomes, after simplifications,

$$\boxed{\ln(1+x) = x - \frac{x^2}{2} + \frac{x^3}{3} + \ldots + (-1)^{n+1}\frac{x^n}{n} + \ldots}. \qquad (3.6.47)$$

⇓

SERIES EXPANSION OF THE LOGARITHM

This expansion was already studied, considering it as a power series associated to the harmonic series. Its domain of convergence is the interval $(-1, 1]$.

Let us take $x = 1$ in this series. It results that

$$\boxed{\ln 2 = 1 - \frac{1}{2} + \frac{1}{3} + \ldots + (-1)^{n+1} \frac{1}{n} + \ldots}. \qquad (3.6.48)$$

Thus, the sum of the alternate harmonic series is $\ln 2$.

* a) LOGARITHMIC TABLES

The logarithm series can help us to calculate logarithms.

Indeed, suppose that we computed $\ln N$, and we wish to estimate $\ln(N+1)$. We calculate the difference

$$D_N = \ln(N+1) - \ln N = \ln \frac{N+1}{N} = \ln\left(1 + \frac{1}{N}\right). \qquad (3.6.49)$$

Taking the series expansion of the logarithm at $x = \frac{1}{N}$, we find

$$D_N = \ln\left(1 + \frac{1}{N}\right) = \frac{1}{N} - \frac{1}{2} \cdot \frac{1}{N^2} + \frac{1}{3} \cdot \frac{1}{N^3} - \frac{1}{4} \cdot \frac{1}{N^4} - \ldots \qquad (3.6.50)$$

Yet, being an alternate series, this series is weakly convergent. But we can improve its convergence.

To do this, we consider the following expansions:

$$\ln(1+x) = x - \frac{x^2}{2} + \frac{x^3}{3} + \ldots + (-1)^{n+1}\frac{x^n}{n} + \ldots,$$
$$\ln(1-x) = -x - \frac{x^2}{2} - \frac{x^3}{3} - \ldots - \frac{x^n}{n} + \ldots .$$
(3.6.51)

Subtracting the second expansions from the first one, we get

$$\ln\frac{1+x}{1-x} = 2\left(x + \frac{x^3}{3} + \frac{x^5}{5} \ldots + \frac{x^{2n-1}}{2n-1} + \ldots\right). \quad (3.6.52)$$

Now, we determine x such that

$$\frac{1+x}{1-x} = 1 + \frac{1}{N} \;\rightarrow\; x = \frac{1}{2N+1}. \quad (3.6.53)$$

By replacing this value of x in (3.6.52), we infer

$$\frac{1}{2}D_N = \frac{1}{2}\ln\left(1 + \frac{1}{N}\right) =$$
$$= \frac{1}{2}\left(\frac{1}{2N+1} + \frac{1}{3}\frac{1}{(2N+1)^3} + \frac{1}{5}\frac{1}{(2N+1)^5} + \ldots \right.$$
$$\left. + \frac{1}{(2n-1)}\frac{1}{(2N+1)^{2n-1}} + \ldots\right), \quad (3.6.54)$$

which is fast convergent and it is commonly used to elaborate logarithmic tables.

* b) EULER'S CONSTANT

Let's apply to function $\ln(1+x)$ Taylor formula, limiting us to the first two terms:

$$\ln(1+x) = x - \frac{x^2}{2}\cdot\frac{1}{(1+\theta x)^2},\quad 0 < \theta < 1. \quad (3.6.55)$$

The last term is the Lagrange remainder R_2, and it tends to 0, by virtue of the series convergence. We can write the formula (3.6.55) in the form

$$x - \ln(1+x) = \frac{x^2}{2} \cdot \frac{1}{(1+\theta x)^2}. \qquad (3.6.56)$$

By taking $x = \frac{1}{n}$, it results that the series of general term $u_n = \frac{1}{n} - \ln\left(1 + \frac{1}{n}\right)$ converges. The sequence of the partial sums

$$A_n = 1 + \frac{1}{2} + \frac{1}{3} + \ldots + \frac{1}{n} - \ln\frac{2}{1} \cdot \frac{3}{2} \cdot \frac{4}{3} \ldots \frac{n+1}{n} =$$
$$= 1 + \frac{1}{2} + \frac{1}{3} + \ldots + \frac{1}{n} - \ln(n+1) \qquad (3.6.57)$$

is obviously also convergent. Denoting by

$$H_n = 1 + \frac{1}{2} + \frac{1}{3} + \ldots + \frac{1}{n} \qquad (3.6.58)$$

the general term of the harmonic series, it follows that

$$A_n = H_n - \ln n + \ln\frac{n}{n+1}. \qquad (3.6.59)$$

As $\lim_{n\to\infty} A_n$ exists, it follows that $\lim_{n\to\infty}(H_n - \ln n)$ also exists.

This limit is called **Euler's constant** and it reads

$$\gamma = 0.5772156\ldots. \qquad (3.6.60)$$

3.7. SEQUENCES AND SERIES OF FUNCTIONS

Consider a sequence whose terms are the functions:

$$\{f_1, f_2, f_3, \ldots, f_n, \ldots\} \equiv \{f_n\}_{n \in \mathfrak{N}}, \qquad (3.7.1)$$

where $f_n : X \subseteq \mathfrak{R} \to \mathfrak{R}$. We suppose that for each $x = x_0 \in X$, the sequence of numbers

$$\{f_1(x_0), f_2(x_0), f_3(x_0), \ldots, f_n(x_0), \ldots\} \equiv$$
$$\equiv \{f_n(x_0)\}_{n \in \mathfrak{N}}, \qquad (3.7.2)$$

has a finite limit: $\lim_{n \to \infty} f_n(x_0) = f_{x_0}$. This way, we set up the correspondence:

$$x \in X \to f_x, \qquad (3.7.3)$$

thus introducing the function $f : X \to \mathfrak{R}, \ f(x) = f_x$.

Definition. The function $f(x) = \lim_{n \to \infty} f_n(x_0), \ x \in X$ is called ***the limit function of the sequence*** $\{f_n\}_{n \in \mathfrak{N}}$.

Examples – in all cases, $X = [0,1]$.

1. The sequence $\{x^n\}_{n \in \mathfrak{N}}$, of general term $f_n(x) = x^n$. Its limit function is

$$f(x) = \begin{cases} 0, & x < 1, \\ 1, & x = 1, \end{cases} \qquad (3.7.4)$$

of graph

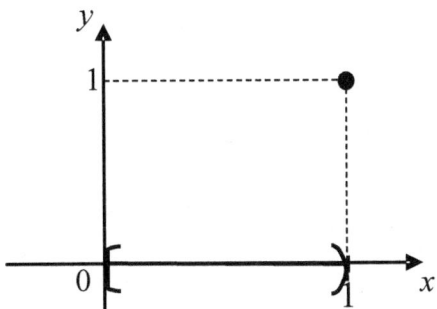

Figure 3.3. The graph of the limit function of the sequemce $\{x^n\}_{n\in\mathfrak{N}}$

2. The sequence $\left\{\dfrac{1}{1+nx}\right\}_{n\in\mathfrak{N}}$, of general term $f_n(x) = \dfrac{1}{1+nx}$. Its limit function is

$$f(x) = \begin{cases} 0, & x > 0, \\ 1, & x = 0, \end{cases} \qquad (3.7.5)$$

having the graph

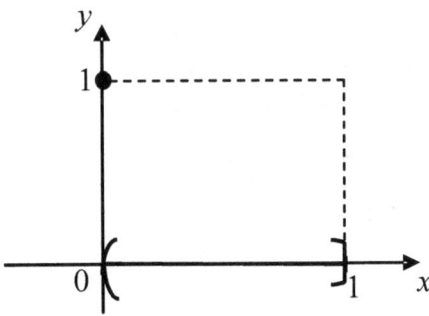

Figure 3.4. The graph of the limit function of the sequence $\left\{\dfrac{1}{1+nx}\right\}_{n\in\mathfrak{N}}$

In these cases, **the limit function is not continuous on** X.

3. The sequence $\left\{\dfrac{x}{1+n^2x^2}\right\}_{n\in\mathfrak{N}}$, of general term $f_n(x) = \dfrac{x}{1+n^2x^2}$. Its limit function is

$$f(x) = 0, \ x \in X, \qquad (3.7.6)$$

having the graph

Figure 3.5. The graph of the linit function of the sequence

$$\left\{\dfrac{x}{1+n^2x^2}\right\}_{n\in\mathfrak{N}}$$

4. The sequence $\left\{2n^2xe^{-n^2x^2}\right\}_{n\in\mathfrak{N}}$, of general term $f_n(x) = 2n^2xe^{-n^2x^2}$. Its limit function is

$$f(x) = 0, \ x \in X, \qquad (3.7.7)$$

i.e., the same function as in the previous example.

Its graph is presented in figure 3.6, which, obviously, coincides with figure 3.5.

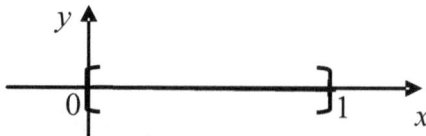

Figure 3.6. The graph of the limit function of the sequence

$$\left\{2n^2xe^{-n^2x^2}\right\}_{n\in\mathfrak{N}}$$

In cases 3 and 4, *the limit function is continuous on* X.

3.7.1. THE SEQUENCE-SERIES EQUIVALENCE

The series study is, in fact, equivalent to the study of sequences; this fact holds true not only in the case of series and sequences of numbers, but also in the more general case, of sequences and series of functions.

1. series ⇒ sequence:

To the series $\sum_{n=1}^{\infty} u_n(x)$, $u_n : X \subseteq \Re \to \Re$, we associate its sequence of partial sums

$$\{f_n\}_{n \in \mathfrak{N}}, \quad f_n(x) = \sum_{n=1}^{n} u_n(x).$$

2. sequence ⇒ series:

To the sequence $\{f_n\}_{n \in \mathfrak{N}}$, we associate the series

$$\sum_{n=1}^{\infty} u_n(x), \quad u_0 = f_1, \quad u_n = f_n - f_{n-1},$$

whose partial sums are $\{f_n\}_{n \in \mathfrak{N}}$, $f_n(x) = \sum_{n=1}^{n} u_n(x)$.

3.7.2. UNIFORM CONVERGENCE, NON-UNIFORM CONVERGENCE

Consider again function sequence $\{f_n\}_{n \in \mathfrak{N}}$ and let f be its limit function. Hence

$$\lim_{n\to\infty} f_n(x) = f(x). \qquad (3.7.8)$$

This means that for any $\varepsilon > 0$ there exists a rank N_ε such that

$$|f_n(x) - f(x)| < \varepsilon, \quad \forall n > N_\varepsilon. \qquad (3.7.9)$$

But N_ε can vary with $x \in X$, therefore $N_\varepsilon = N_\varepsilon(x)$.

- If N_ε is the same for each $x \in X$, then the **convergence is uniform**.
- If $N_\varepsilon = N_\varepsilon(x)$, then the **convegence is not uniform**.

WHEN DOES N_ε DEPEND ON $x \in X$?

We have pro & con examples:

- Take the sequence of general term

$$f_n(x) = \frac{x}{1 + n^2 x^2}. \qquad (3.7.10)$$

We already saw that

$$\lim_{n\to\infty} f_n(x) = f(x) = 0, \quad x \in [0,1],$$

so the limit function f is identically zero. Obviously,

$$0 \le f_n(x) = \frac{1}{2n}\frac{2nx}{1 + n^2 x^2} < \frac{1}{2n}, \quad x \in [0,1].$$

Consequently, if we take $N_\varepsilon = \left[\dfrac{1}{2\varepsilon}\right] + 1$, then

$$0 \le f_n(x) < \varepsilon, \ x \in [0,1].$$

In this case, N_ε does not depend on x, so the sequence **converges uniformly** to its limit function $f \equiv 0$.

- Now, consider the sequence of general term

$$f_n = x^n, \ x \in [0,1]. \quad (3.7.11)$$

From figure 3.7, we see that the convergence is loosing uniformity, by and by, as n increases. Indeed, if $\varepsilon < 1$, then it is impossible that $x^n < \varepsilon$, $\forall x \in [0,1]$, because, if $x \to 1$, then $x^n \to 1$, for n fixed.

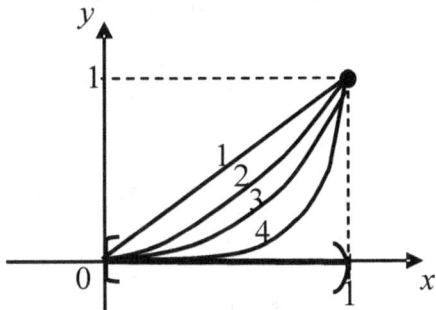

Figure 3.7. Non-uniform convergence

3.7.3. WEIERSTRASS CRITERION

Theorem 3.9 (WEIERSTRASS). *Consider the series* $\sum_{n=1}^{\infty} u_n(x)$, $u_n : X \subseteq \Re \to \Re$. *If*

$$|u_n(x)| \le c_n, \ \forall n \in \mathfrak{N}, \forall x \in X, \quad (3.7.12)$$

and $\sum_{n=1}^{\infty} c_n$ is convergent, then $\sum_{n=1}^{\infty} u_n(x)$ **is uniformly and absolutely convergent on X.**

***Proof.** The series $\sum_{n=1}^{\infty} c_n$ is convergent. Take $\varepsilon > 0$. Then, according to the fundamental theorem for number series, there exists N_ε such that

$$0 < R_{np} \equiv c_{n+1} + c_{n+2} + \ldots + c_{n+p} < \varepsilon, \forall n > N_\varepsilon, p \in \mathfrak{N}. \quad (3.7.13)$$

It is important to mention that N_ε **does not depend on** x.

From (3.7.12) it follows that the finite remainder of the series $\sum_{n=1}^{\infty} |u_n(x)|$ also satisfies the inequality

$$|u_{n+1}(x)| + |u_{n+2}(x)| + \ldots + |u_{n+p}(x)| < \varepsilon, \\ \forall n > N_\varepsilon, p \in \mathfrak{N}, \quad (3.7.14)$$

for any $x \in X$, and N_ε does not depend on x!

It follows that the series is absolutely and uniformly convergent on X. ∎

Example. The series $\sum_{n=1}^{\infty} a_n \cos nx$, $\sum_{n=1}^{\infty} b_n \sin nx$ are uniformly convergent on any real interval, if the numerical series $\sum_{n=1}^{\infty} |a_n|$, $\sum_{n=1}^{\infty} |b_n|$ are convergent.

Indeed, we have

$$|a_n \cos nx| \leq |a_n|, \quad |b_n \sin nx| \leq |b_n|, \quad \forall x \in \Re. \qquad (3.7.15)$$

3.7.4. PROPERTIES OF THE SUMS OF UNIFORMLY CONVERGENT SERIES OF FUNCTIONS

We give, without proof, some important properties of the sums of uniformly convergent series.

1. ***The sum of a uniformly convergent series of continuous functions is a continuous function.***

   Example

- A power series is ***continuous*** on its entire domain of convergence.
- The series

$$(1-x) + x(1-x) + x^2(1-x) + \ldots + x^n(1-x) + \ldots$$

is absolutely convergent for $|x| < 1$ and also converges at $x = 1$. The sum of the series is the limit function of the partial sum sequence, hence

$$f(x) = \begin{cases} (1-x) \cdot \dfrac{1}{(1-x)} = 1, & |x| < 1, \\ 0, & x = 1. \end{cases} \qquad (3.7.16)$$

The sum of the series is a ***discontinuous*** function, although its terms are all of them ***continuous***. The reason for this mismatch is that the ***series is not uniformly convergent***.

2. A convergent series of differentiable functions can be differentiated term by term, if the series of its derivatives is uniformly convergent:

$$\left(\sum_{n=1}^{\infty} u_n(x)\right)' = \sum_{n=1}^{\infty} u'_n(x). \qquad (3.7.17)$$

Examples

a) Can the following series be differentiated term by term

$$\sin x + \frac{1}{2^2}\sin\frac{x}{2} + \frac{1}{3^2}\sin\frac{x}{3} + \ldots + \frac{1}{n^2}\sin\frac{x}{n} + \ldots ?$$

Solution. The series is convergent (even absolutely and uniformly) on any real interval, according to Weierstrass criterion. Indeed, we have

$$\frac{1}{n^2}\left|\sin\frac{x}{n}\right| \leq \frac{1}{n^2},$$

hence its general term is majoreized by the general term of the convergent Riemman series

$$1 + \frac{1}{2^2} + \frac{1}{3^2} + \ldots + \frac{1}{n^2} + \ldots .$$

Also, the series of the derivatives

$$\cos x + \frac{1}{2^3}\cos\frac{x}{2} + \frac{1}{3^3}\cos\frac{x}{3} + \ldots + \frac{1}{n^3}\cos\frac{x}{n} + \ldots$$

is uniformly convergent on any real interval, again due to Weierstrass criterion, because of the inequality

$$\frac{1}{n^3}\left|\cos\frac{x}{n}\right| \le \frac{1}{n^3}.$$

Consequently, *the series can be differentiated term by term.*

b) Can the following series be differentiated term by

$$\frac{\cos x}{1} + \frac{\cos 2x}{2} + \frac{\cos 3x}{3} + \ldots + \frac{\cos nx}{n} + \ldots?$$

Solution. The series is convergent (for ex., on the interval $\left[\frac{\pi}{4}, \frac{3\pi}{4}\right]$, according to the generalized Dirichlet criterion for series of functions), but the series of derivatives

$$\sin x + \sin 2x + \ldots + \sin nx + \ldots$$

is not uniformly convergent.

So *the series cannot be differentiated term by term.*

3. *A uniformly continuous series of continuous functions on an interval* $[a,b] \subset \infty$ *can be integrated term by term:*

$$\int_a^b \left(\sum_{n=1}^{\infty} u_n(x)\right) dx = \sum_{n=1}^{\infty} \int_a^b u_n(x) dx. \qquad (3.7.18)$$

Example. Can the following series

$$\cos x + \frac{1}{2}\cos 2x + \frac{1}{2^2}\cos 3x + \ldots + \frac{1}{2^{n-1}}\cos nx + \ldots$$

be integrated on the interval $\left[\frac{\pi}{4}, \frac{\pi}{3}\right]$?

Solution. The terms of the series are continuous, and the series is uniformly convergent on any real interval, according to Weierstrass criterion, because

$$\frac{1}{2^{n-1}}|\cos nx| \leq \frac{1}{2^{n-1}};$$

this inequality means that its general term is majorized by the general term of the convergent geometric progression

$$1 + \frac{1}{2} + \frac{1}{2^2} + \ldots + \frac{1}{2^{n-1}} + \ldots.$$

Hence, *the series can be integrated term by term* on $\left[\frac{\pi}{4}, \frac{3\pi}{4}\right]$.

3.7.5. POWER SERIES (RESUMPTION)

We saw that a function series can be differentiated term by term if the series of the derivatives is uniformly convergent. Let us take the particular case of power series. Therefore, consider the series

I. $f(x) = a_0 + a_1 x + a_2 x^2 + \ldots + a_n x^n + \ldots \quad \to R,$

II. $f_1(x) = a_1 + 2a_2 x + 3a_3 x^2 + \ldots + n a_n x^{n-1} + \ldots \to R_1,$ (3.7.19)

having the convergence radiuss R, respectively R_1.

We will show that both convergence radii coincide.

* **Proof.**

♣ Suppose that $\boxed{R_1 > R}$.

Then there exists $\rho > 0$, $R < \rho < R_1$, such that $f_1(x)$ be absolutely convergent for $|x| < R_1$, which means that the series

$$S \equiv |a_1| + 2|a_2|\rho + 3|a_3|\rho^2 + \ldots + n|a_n|\rho^{n-1} + \ldots \quad (3.7.20)$$

is convergent. From here, it easily follows that the series with positive terms

$$\begin{aligned}S\rho + |a_0| &\equiv \\ &\equiv |a_0| + |a_1|\rho + 2|a_2|\rho^2 + 3|a_3|\rho^3 + \ldots + n|a_n|\rho^n + \ldots\end{aligned} \quad (3.7.21)$$

is also convergent.

But, obviously, $|a_n|\rho^n \leq n|a_n|\rho^n$. The modulus of the general term of $f(\rho)$ is thus majorized by the general term of a convergent series. By virtue of comparison theorem, it follows that $\sum_{n=0}^{\infty} |a_n|\rho^n$ is also convergent. But $\rho > R$, which is contradictory. It results that $R_1 \leq R$.

♣ Now, let $\boxed{R_1 < R}$.

Then, there exists $\rho > 0$, strictly included in the interval (R_1, R). The series **I** is absolutely convergent for $|x| < R$, so its general term is at least bounded at ρ, i.e.

$$|a_n|\rho^n < M \Rightarrow |a_n| < \frac{M}{\rho^n}, \quad n \in \mathfrak{N}. \quad (3.7.22)$$

Let $R_1 < x < \rho < R$, and consider the series $\sum_{n=1}^{\infty} n|a_n||x|^{n-1}$. Its general term is majorized by:

$$n|a_n||x|^{n-1} = n|a_n|\left|\frac{x}{\rho}\right|^{n-1} \rho^{n-1} < \frac{M}{\rho^n} n \left|\frac{x}{\rho}\right|^{n-1} \rho^{n-1}, \qquad (3.7.23)$$

hence

$$n|a_n||x|^{n-1} < \frac{M}{\rho} nr^{n-1}, \quad r = \left|\frac{x}{\rho}\right|^{n-1} < 1. \qquad (3.7.24)$$

On the right side of this inequality, one finds the general term of the series

$$\frac{M}{\rho}\left(1 + 2r + 3r^2 + \ldots + nr^{n-1} + \ldots\right), \qquad (3.7.25)$$

which converges for $r < 1$. Its sum is $\frac{M}{\rho} \cdot (1-r)^{-2}$, and it was obtained by means of the generalized binomial series. Consequently, the series **II** would be absolutely convergent for $x > R_1$, which is contradictory. It follows that $R = R_1$.

From the above considerations, we see that

A power series can be differentiated term by term on its domain of convergence.

If

$$f(x) = a_0 + a_1 x + a_2 x^2 + \ldots + a_n x^n + \ldots, \qquad (3.7.26)$$

then

$$f'(x) = a_1 + 2a_2 x + 3a_3 x^2 + \ldots + na_n x^{n-1} + \ldots, \qquad (3.7.27)$$

with the same radius of convergence as f. Differentiating again, we obtain

$$f''(x) = 2 \cdot 1 a_2 + 3 \cdot 2 a_3 x + \ldots + n(n-1) a_n x^{n-2} + \ldots . \quad (3.7.28)$$

Hence, the derivative of an arbitrary order k of f is

$$f^{(k)}(x) = \\ = k! a_k + \frac{(k+1)!}{1} a_{k+1} x + \frac{(k+2)!}{1 \cdot 2} a_{k+2} x^2 + \ldots . \quad (3.7.29)$$

Taking the series at $x = 0$, the coefficients a_k read

$$a_0 = f(0), \quad a_1 = \frac{f'(0)}{1!}, \quad a_2 = \frac{f''(0)}{2!}, \ldots, \\ a_k = \frac{f^{(k)}(0)}{k!}, \ldots . \quad (3.7.30)$$

By replacing these values in the expression of $f(x)$, we obtain

$$f(x) = f(0) + \frac{f'(0)}{1!} x + \frac{f''(0)}{2!} x^2 + \ldots + \\ + \frac{f^{(k)}(0)}{k!} x^k \ldots, \quad (3.7.31)$$

which coincides with the development in Mac Laurin series of $f(x)$. It results the following important fact:

The power series expansion of a function is identical with its Mac Laurin series expansion.

A function allowing such an expansion is usually called **analytic function**.

3.7.5.1. Applications

1. Calculate the sum of the series

$$f(x) = \frac{x}{1} - \frac{x^2}{2} + \frac{x^3}{3} - \ldots .$$

Solution. The radius of convergence of the series can be determined by D'Alembert (or ratio) test:

$$\lim_{n\to\infty}\left|\frac{a_{n+1}}{a_n}\right| = \lim_{n\to\infty}\frac{\frac{1}{n+1}}{\frac{1}{n}} = \lim_{n\to\infty}\frac{n}{n+1} = 1.$$

It coincides with the radius of convergence of the series of derivatives:

$$f'(x) = 1 - x + x^2 - x^3 + \ldots = \frac{1}{1+x}.$$

It follows that $f(x) = \ln(1+x) + C$ and, as $f(0) = 0$, we finally obtain $f(x) = \ln(1+x)$, therefore

$$\boxed{\ln(1+x) = \frac{x}{1} - \frac{x^2}{2} + \frac{x^3}{3} - \ldots},$$

which we already knew.

2. Compute the sum of the series

$$f(x) = \frac{x}{1} - \frac{x^3}{3} + \frac{x^5}{5} - \ldots .$$

Solution. The radius of convergence of the series can be determined by using D'Alembert's test:

$$\lim_{n\to\infty}\left|\frac{a_{n+1}}{a_n}\right| = \lim_{n\to\infty}\frac{\frac{1}{2n+1}}{\frac{1}{2n-1}} = \lim_{n\to\infty}\frac{2n-1}{2n+1} = 1.$$

It coincides with the radius of convergence of the series of derivatives:

$$f'(x) = 1 - x^2 + x^4 - x^6 + \ldots = \frac{1}{1+x^2}.$$

It follows that $f(x) = \arctan x + C$ and, as $f(0) = 0$, we finally obtain $f(x) = \arctan x$, hence

$$\boxed{\arctan x = \frac{x}{1} - \frac{x^3}{3} + \frac{x^5}{5} - \frac{x^7}{7}\ldots}.$$

Taking $x = 1$ in this power series, we find

$$\boxed{\frac{\pi}{4} = 1 - \frac{1}{3} + \frac{1}{5} - \frac{1}{7}\ldots},$$

an expansion that can be used to approximate π.

3.7.6. COMPUTING LIMITS OF FUNCTIONS AND DEFINITE INTEGRALS BY MEANS OF POWER SERIES

Using power series, one can put into evidence standard procedures for the calculus of definite integrals and of limits of functions, which, otherwise calculated (with classical methods), require special skills and/or laborious computation. We give below two such examples, the first one dealing with the calculus of limits of functions and the second one – with the approximation of a definite integral.

1. $L = \lim\limits_{x \to 0} \dfrac{2e^x - 2 - 2x - x^2}{x - \sin x}$.

We use the power series expansion of the exponential and of the sinus around $x = 0$:

$$L = \lim_{x \to 0} \dfrac{2\left(1 + \dfrac{x}{1!} + \dfrac{x^2}{2!} + \dfrac{x^3}{3!} + \ldots\right) - 2 - 2x - x^2}{x - \left(x - \dfrac{x^3}{3!} + \dfrac{x^5}{5!} - \ldots\right)} =$$

$$= \lim_{x \to 0} \dfrac{\dfrac{2}{3!} + \dfrac{2x}{4!} + \ldots}{\dfrac{1}{3!} - \dfrac{x^2}{5!} + \ldots},$$

hence

$$\boxed{\lim_{x \to 0} \dfrac{2e^x - 2 - 2x - x^2}{x - \sin x} = 2}.$$

2. We write the power series expansion of the logarithm.

$$\int_0^{0.1} \dfrac{\ln(1+x)}{x} dx = \int_0^{0.1} \dfrac{1}{x}\left(x - \dfrac{x^2}{2} + \dfrac{x^3}{3} - \dfrac{x^4}{4} + \ldots\right) dx =$$

$$= \int_0^{0.1} \left(1 - \dfrac{x}{2} + \dfrac{x^2}{3} - \dfrac{x^3}{4} + \ldots\right) dx =$$

$$= \left(x - \dfrac{x^2}{4} + \dfrac{x^3}{9} - \dfrac{x^4}{16} + \ldots\right)\Big|_0^{0.1} =$$

$$= 0.1 - \dfrac{0.01}{4} + \dfrac{0.001}{9} - \dfrac{0.0001}{16} + \ldots,$$

therefore

$$\boxed{\int_0^{0.1} \frac{\ln(1+x)}{x} dx \cong 0.098}.$$

3.7.7. POWER SERIES IN COMPLEX

These are series of the form

$$\sum_{n=0}^{\infty} a_n (z - z_0)^n,$$

where $a_n = \alpha_n + i\beta_n$, and z, z_0 are also complex. With the notation $\zeta = z - z_0$, such series read

$$\sum_{n=0}^{\infty} a_n \zeta^n,$$

hence, the study of power series in complex is reduced to the study of series of the simpler form

$$\sum_{n=0}^{\infty} a_n z^n.$$

Abel (theorem 3.1) and Cauchy-Hadamard (theorem 3.2) theorems, previously stated and proved, allow natural extensions in complex.

Theorem 3.10.(ABEL) *Given a power series* $\sum_{n=0}^{\infty} a_n z^n$, *one can always find a positive real number R called radius of convergence, to which a circle* $|z| = R$ *corresponds in the* complex plane, called *circle of convergence*, with the following properties:

1) in the disk $|z| < R$ (interior of the circle) the series is convergent;

2) in exterior, for $|z| > R$, the series does not converge;

3) in any interior disk $|z| \leq \delta < R$ the series is uniformly and absolutely convergent.

The radius of convergence is calculated via Cauchy-Hadamard theorem, in the same way as in the real case.

Many of the properties of power series, some of them previously mentioned, are also valid in complex.

For example, the series $\sum_{n=1}^{\infty} n a_n z^{n-1}$ has the same radius of convergence as the series $\sum_{n=0}^{\infty} a_n z^n$, and its sum is the derivative of the sum of the series $\sum_{n=0}^{\infty} a_n z^n$.

Example. Calculate the radius of convergence of the power series

$$S(z) \equiv 1 + z + \frac{z^2}{2!} + \frac{z^3}{3!} + \ldots + \frac{z^n}{n!} + \ldots .$$

According to lemma 3.1, we have

$$R = \frac{1}{\lim_{n \to \infty} \frac{n!}{(n+1)!}} = \infty ,$$

so the series is convergent for all z in the complex plane (except, of course, the point at infinity). The derivative of S verifies, obviously, the equation $S'(z) = S(z)$, so $S(z) = c e^z$. Because $S(0) = 1$, it follows that $c = 1$, hence $S(z) = e^z$.

Let us remember that we already established Euler's formulas. We have

$$e^x (\cos y + i \sin y) = e^x \cdot e^{iy} = e^{x+iy},$$

therefore

$$e^z = e^x (\cos y + i \sin y),$$

a formula very useful in applications.

3.8. FOURIER SERIES

Joseph Fourier, the illustrious French mathematician, thought that the fundamental purpose of Mathematics, as science, is to understand natural phenomena, in order to foresee and apply them. However, his scientific opera did not emphasize the practical advantages of Mathematics.

The one who proved how strong and useful was the mathematical tool created by Fourier was William Thompson, lord Kelvin, another great personality of that era, an eminent physician.

Following Körner [6], we will present here one of the most simple ideas of Fourier, put into practice by Kelvin.

Mariner's compass. Consider the problem of correcting a magnetic compass, mounted on a ship which contains a big quantity of iron and steel. Suppose that the magnetic compass indicates an angle θ with the North. But the true angle is in fact $f(\theta) = \theta + \varepsilon(\theta)$, where $\varepsilon(\theta)$ is the error. The factors responsible for this error are mainly:

- the permanent magnetism of the ship, produced by the earth's magnetic field affecting the ship's hard iron during construction;
- the magnetism induced by the magnetic terrestrial field.

Hence, computing $\varepsilon(\theta)$ is not an easy task. Its value can be big enough – even up to 20^0 – but it suffices to compute it with a precision of only 2^0 or 3^0, taking into account that there are other physical errors in setting a course which have the same order of magnitude.

Therefore, it makes sense to approximate $\varepsilon(\theta)$ by a low degree trigonometric polynomial, say

$$\varepsilon_1(\theta) = a_0 + (a_1 \cos\theta + b_1 \sin\theta) + (a_2 \cos 2\theta + b_2 \sin 2\theta).$$

The values of the coefficients can be computed by moving the ship round in port, so that its direction θ with respect to known landmarks is thoroughly measured, meanwhile noting the values of $f(\theta)$ for various θ.

This was the state-of-art in this problem when, in 1871, Kelvin was asked by a magazine to write an article on mariner's compass. He realized very quickly that he knew almost nothing of the subject. It took him three years to write the first article and five more years to write the second and to redesign the magnetic compass. Kelvin reduced its size, such that the compass could be corrected directly, by the appropriate positionning of permanent magnets and spheres of iron close to it.

Epilogue. After the customary Victorian battle with the Admiralty, the compass and the method of correction suggested by Kelvin were adopted and they remained in use up to The Second World War.

This episode from the history of science fully justifies a careful study of Fourier series.

3.8.1. PERIODIC FUNCTIONS

The periodic functions have been studied, and still are, for their practical importance.

Definition. A real function $f : \Re \to \Re$ is called *periodic* if a real number $T \neq 0$ exists such that $f(x+T) = f(x)$, $\forall x \in \Re$.

T is called *period*.

Obviously, if T is period, then nT, for any n natural, is also period.

Any non-constant periodic function with at least one point of continuity allows a least positive period. In fact, this least period is what we usually call a period.

The most natural properties of periodic functions are:

- If f has the period T, then $\varphi(x) = f(ax)$ has the period $\dfrac{T}{a}$.

This can be verified directly. We have

$$\varphi\left(x + \frac{T}{a}\right) = f(ax + T) = f(ax) = \varphi(x). \qquad (3.8.1)$$

- If f (integrable) is periodic, then the following integral relation is valid

$$\int_c^{c+T} f(x)\,dx = \int_0^T f(x)\,dx, \qquad (3.8.2)$$

for arbitrary $c \in \Re$.

Indeed,

$$\int_c^{c+T} f(x)\,dx = \int_c^0 f(x)\,dx + \int_0^T f(x)\,dx + \int_T^{c+T} f(x)\,dx =$$

$$= -\int_0^c f(x)\,dx + \int_0^T f(x)\,dx + \int_0^c f(u+T)\,du = \qquad (3.8.3)$$

$$= -\int_0^c f(x)\,dx + \int_0^T f(x)\,dx + \int_0^c f(u)\,du = \int_0^T f(x)\,dx.$$

3.8.2. FOURIER SERIES

The simplest and best known periodic functions are the trigonometric functions. It is thus natural to consider the problem of decomposing a periodic function in simple harmonics, i.e.

$$a\cos\omega x + b\sin\omega x;$$

in this case, the period is $2\pi/\omega$. If f is a periodic function with period 2π, then the frequencies of its harmonic components must be chosen such that $n \cdot \frac{2\pi}{\omega} = 2\pi$, hence $n = \omega$. So, the decomposition problem reduces to the following

PROBLEM. *Expand a periodic function (with period 2π) in a series of the form*

$$\frac{a_0}{2} + (a_1 \cos x + b_1 \sin x) + (a_2 \cos 2x + b_2 \sin 2x) +$$
$$+ \ldots + (a_n \cos nx + b_n \sin nx) + \ldots \equiv \qquad (3.8.4)$$
$$\equiv \frac{a_0}{2} + \sum_{n=1}^{\infty}(a_n \cos nx + b_n \sin nx),$$

where a_0, a_n, b_n, $n \in \mathcal{N}$, are constants that must be determined.

Suppose that we have found for $f(x)$ a uniformly convergent series of the form

$$f(x) = \frac{a_0}{2} + \sum_{n=1}^{\infty}(a_n \cos nx + b_n \sin nx). \qquad (3.8.5)$$

Being uniformly convergent, this series can be integrated term by term, from $-\pi$ to π, and it results that:

$$\int_{-\pi}^{\pi} f(x)dx =$$

$$= \frac{a_0}{2} \int_{-\pi}^{\pi} dx + \sum_{n=1}^{\infty} a_n \int_{-\pi}^{\pi} \cos nx dx + \sum_{n=1}^{\infty} b_n \int_{-\pi}^{\pi} \sin nx dx,$$

wherefrom

$$a_0 = \frac{1}{\pi} \int_{-\pi}^{\pi} f(x)dx. \qquad (3.8.6)$$

Further, by multiplying (3.8.5) with $\cos kx$ and integrating again, we deduce that

$$a_k = \frac{1}{\pi} \int_{-\pi}^{\pi} f(x)\cos kx dx, \quad k \in \mathcal{N}, \qquad (3.8.7)$$

because

$$\int_{-\pi}^{\pi} \cos nx \cos kx dx = \begin{cases} 0, & n \neq k, \\ \pi, & n = k, \end{cases}$$

$$\int_{-\pi}^{\pi} \sin nx \cos kx dx = 0. \qquad (3.8.8)$$

Now, by multiplying (3.8.5) with $\sin kx$ and integrating the result on $[-\pi, \pi]$, we find:

$$b_k = \frac{1}{\pi} \int_{-\pi}^{\pi} f(x)\sin kx dx, \quad k \in \mathcal{N}, \qquad (3.8.9)$$

because

$$\int_{-\pi}^{\pi} \sin nx \sin kx dx = \begin{cases} 0, & n \neq k, \\ \pi, & n = k. \end{cases} \qquad (3.8.10)$$

With these preparations, we can introduce the following

Definition. Let f be a periodic function with period 2π, absolutely integrable on $[-\pi, \pi]$, and having on $[-\pi, \pi]$ a finite number of discontinuity points of first kind. ***The Fourier series*** associated to f is the series (3.8.5), whose coefficients a_0, a_k, b_k, $k \in \mathcal{N}$, are given by formulas (3.8.6), (3.8.7) and (3.8.9).

A periodic function can be expanded in a Fourier series at any derivability point. At these points, the series converges to the corresponding value of f.

A standard class of functions expandable in a Fourier series is the class of functions with the Dirichlet property.

Definition. We say that f has the ***Dirichlet property*** on the interval $[a,b]$ if this interval can be divided in a finite number of subintervals such that the function be monotonic and bounded on each of them (figure 3.8).

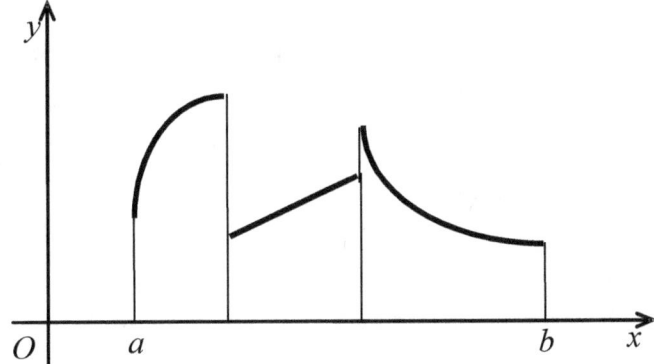

Figure 3.8. An example of function with Dirichlet property

We give the following theorem without proof:

Theorem 3.11. *If the function $f: \Re \to \Re$, periodic and with period T, has the Dirichlet property on an interval $[\alpha, \alpha + T]$, then its Fourier series is convergent for any t. The sum $S(t)$ of the Fourier series is equal to $f(t)$ at each point at which $f(t)$ is continuous. At a point of discontinuity c of first kind, $S(c)$ is equal to the arithmetic mean of the two one-sided limits of $f(t)$ at c, i.e.*

$$S(t) = \begin{cases} f(t), & t \text{ continuity point,} \\ \dfrac{f(t_+) + f(t_-)}{2}, & t \text{ discontinuity point.} \end{cases} \quad (3.8.11)$$

The theorem presented above is also called the **Dirichlet theorem** and it constitutes a criterion of wide applicability, when it comes to expanding a function in a Fourier series.

Example. Determine the Fourier series of function ψ, with period 2π, defined on $[0, 2\pi]$ as follows:

$$\psi(x) = \begin{cases} \dfrac{x}{2}, & x \in (0, 2\pi), \\ \dfrac{\pi}{2}, & x \in \{0, 2\pi\}. \end{cases}$$

The graph of this function is drown below, in figure 3.9.

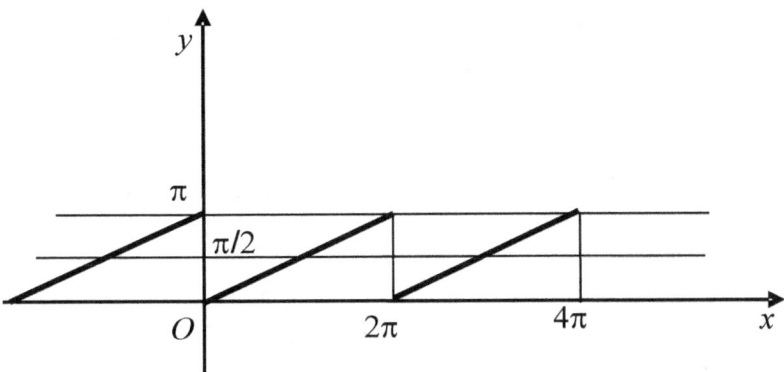

Figure 3.9. The graph of the function ψ

By formula (3.8.6), we have

$$a_0 = \frac{1}{\pi}\int_0^{2\pi} \frac{x}{2}\,dx = \pi.$$

Then

$$a_n = \frac{1}{\pi}\int_0^{2\pi} \frac{x}{2}\cos nx\,dx =$$

$$= \frac{1}{\pi}\left(\frac{x}{2n}\sin nx\right)\Big|_0^{2\pi} - \frac{1}{\pi}\int_0^{2\pi} \frac{\sin nx}{n}\cdot\frac{1}{2}\,dx = 0,$$

and

$$b_n = \frac{1}{\pi}\int_0^{2\pi} \frac{x}{2}\sin nx\,dx =$$

$$= \frac{1}{\pi}\left[\frac{x}{2}\cdot\left(-\frac{\cos nx}{n}\right)\right]\Big|_0^{2\pi} + \frac{1}{\pi n}\int_0^{2\pi}\frac{1}{2}\cos nx\,dx =$$

$$= -\frac{1}{2\pi}\cdot 2\pi\cdot\frac{1}{n} = -\frac{1}{n}.$$

Hence,

$$\psi(x) = \frac{\pi}{2} - \sum_{n=1}^{\infty}\frac{\sin nx}{n},\ x\in\Re,$$

which represents the Fourier series we were looking for.

Using integration by parts, we can find the Fourier series of the derivative of a periodic function. Also, by applying Euler's formulas, we can establish the complex form of a Fourier series, i.e.,

$$S \equiv \frac{a_0}{2} + \sum_{n=1}^{\infty}(a_n \cos nx + b_n \sin nx) =$$
$$= \frac{a_0}{2} + \sum_{n=1}^{\infty} a_n \cdot \frac{e^{inx} + e^{-inx}}{2} + \sum_{n=1}^{\infty} b_n \cdot \frac{e^{inx} - e^{-inx}}{2i} = \quad (3.8.12)$$
$$= \frac{a_0}{2} + \sum_{n=1}^{\infty} \frac{a_n - ib_n}{2} e^{inx} + \sum_{n=1}^{\infty} \frac{a_n - ib_n}{2} e^{-inx}.$$

With the notations

$$c_k = \frac{a_k - ib_k}{2}, \quad c_{-k} = \frac{a_k + ib_k}{2},$$
$$c_0 = \frac{a_0}{2}, \quad (3.8.13)$$

the partial sums S_n of the series S read

$$S_n = c_0 + \sum_{k=1}^{n} c_k e^{ikx} + \sum_{k=1}^{n} c_{-k} e^{-ikx} = \sum_{k=-n}^{n} c_k e^{ikx}. \quad (3.8.14)$$

For the coefficients, it follows that

$$c_k = \frac{a_k - ib_k}{2} =$$
$$= \frac{1}{2\pi}\left(\int_{-\pi}^{\pi} f(x)\cos kx\, dx - i\int_{-\pi}^{\pi} f(x)\sin kx\, dx\right), \quad (3.8.15)$$

which yields

$$c_k = \frac{1}{2\pi} \int_{-\pi}^{\pi} f(x)e^{-ikx}dx, \quad k \in \mathfrak{N}. \qquad (3.8.16)$$

As $c_{-k} = \overline{c_k}$, we also have

$$c_{-k} = \frac{1}{2\pi} \int_{-\pi}^{\pi} f(x)e^{ikx}dx, \quad k \in \mathfrak{N};$$

the coefficient c_0 can be also expressed as

$$c_0 = \frac{1}{2\pi} \int_{-\pi}^{\pi} f(x)e^{i \cdot 0 \cdot x}dx,$$

such that we can finally write the coefficients in the form:

$$c_k = \frac{1}{2\pi} \int_{-\pi}^{\pi} f(x)e^{-ikx}dx, \quad k \in \mathfrak{Z}. \qquad (3.8.17)$$

At the points of derivability of f, we have

$$f(x) = \lim_{n \to \infty} S_n \equiv \lim_{n \to \infty}\left[\frac{a_0}{2} + \sum_{k=1}^{n}(a_k \cos kx + b_k \sin kx)\right] =$$

$$= \lim_{n \to \infty} \sum_{k=-n}^{n} c_k e^{ikx},$$

i.e.,

$$f(x) = \sum_{k=-\infty}^{+\infty} c_k e^{ikx}. \qquad (3.8.18)$$

This formula is ***the complex form of a Fourier series***.

3.8.3. FOURIER SERIES OF ODD AND EVEN FUNCTIONS

It is known that a real function $f : \Re \to \Re$ is called *even* if

$$f(-x) = f(x), \quad x \in \Re, \quad (3.8.19)$$

and *odd* if

$$f(-x) = -f(x), \quad x \subset \Re. \quad (3.8.20)$$

For integrable even functions, we always have

$$\int_{-c}^{c} f(x)dx = 2\int_{0}^{c} f(x)dx; \quad (3.8.21)$$

for the odd ones, we can show that

$$\int_{-c}^{c} f(x)dx = 0. \quad (3.8.22)$$

♣ The Fourier series associated to a *periodic even function* f is

$$f(x) = \frac{a_0}{2} + \sum_{n=1}^{\infty} a_n \cos nx, \quad a_n = \frac{2}{\pi}\int_{0}^{\pi} f(x)\cos nx\, dx, \quad (3.8.23)$$

$$n \in \mathfrak{N}^*,$$

because the coefficients b_n are all of them null by virtue of (3.8.22) and for a_n we have (3.8.21).

♣ The Fourier series associated to a *periodic odd function* is

$$f(x) = \sum_{n=1}^{\infty} b_n \sin nx, \quad b_n = \frac{2}{\pi}\int_0^{\pi} f(x)\sin nx\, dx \qquad (3.8.24)$$

$n \in \mathcal{N},$

because a_0, a_n are all of them null by (3.8.22) and this time we apply (3.8.21) to compute b_n.

Example. Find the Fourier series for the periodic function, defined on $[-\pi, \pi]$ by

$$f(x) = \begin{cases} -\dfrac{\pi}{4}, & x \in (-\pi, 0), \\ \dfrac{\pi}{4}, & x \in (0, \pi), \\ 0, & x \in \{n\pi, n \in \mathcal{Z}\}. \end{cases}$$

The graph of this function is presented in figure 3.10.

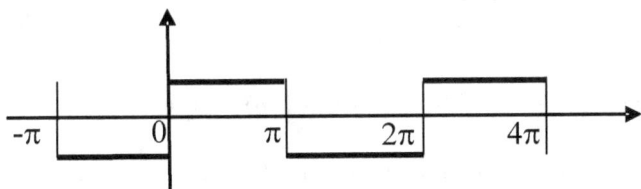

Figure 3.10. The graph of the function f

It is easily seen that this is an odd function, so its Fourier series is a series of sinuses.

As a_0, a_n are all of them null, we must compute only the coefficients b_n. We have

$$b_n = \frac{2}{\pi}\int_0^\pi f(x)\sin nx\, dx = \frac{2}{\pi}\cdot\frac{\pi}{4}\int_0^\pi \sin nx\, dx =$$

$$= \frac{1}{2}\left(-\frac{\cos nx}{n}\right)\Big|_{x=0}^{x=\pi} = \frac{1}{2n}\left[1-(-1)^n\right],$$

which finally gives

$$b_n = \begin{cases} 0, & n = 2m, \\ \dfrac{1}{n}, & n = 2m-1. \end{cases}$$

The desired expansion is :

$$f(x) = \sin x + \frac{1}{3}\sin 3x + \ldots + \frac{1}{2m-1}\sin(2m-1)x + \ldots .$$

This development remains valid also at the points of discontinuity of f. At $\pi/2$, we have

$$f\left(\frac{\pi}{2}\right) = \frac{\pi}{4} = 1 - \frac{1}{3} + \frac{1}{5} - \ldots .$$

This formula can serve to approximate π.

A periodic function with an arbitrary period can be converted in a function with period 2π by a convenient change of variable. Indeed, by appealing the first property of periodic functions, we observe that if $f(x)$ is with period $2l$, as a function of x, then $f\left(\dfrac{lt}{\pi}\right)$ is with period 2π, as a function of t.

So, if f is absolutely integrable on $(-l, l)$ and has a finite number of points of discontinuity, then

$$f\left(\frac{lt}{\pi}\right) = \frac{a_0}{2} + \sum_{n=1}^{\infty}(a_n \cos nt + b_n \sin nt),$$

where the expressions of the coefficients are given by formulas (3.8.6), (3.8.7) and (3.8.9). To get back to the variable x, we resume these formulas:

$$a_0 = \frac{1}{\pi}\int_{-\pi}^{\pi} f(\frac{lt}{\pi})dt = \frac{1}{\pi}\int_{-l}^{l} f(x) \cdot \frac{\pi}{l}dx = $$
$$= \frac{1}{l}\int_{-l}^{l} f(x)dx.$$
(3.8.25)

By proceeding in the same way with the other formulas, we deduce

$$a_n = \frac{1}{l}\int_{-l}^{l} f(x)\cos\frac{n\pi x}{l}dx,$$
$$b_n = \frac{1}{l}\int_{-l}^{l} f(x)\sin\frac{n\pi x}{l}dx, \quad n \in \mathfrak{N}.$$
(3.8.26)

In conclusion, we obtain the Fourier series of a periodic function, of period $2l$, as

$$f(x) = \frac{a_0}{2} + \sum_{n=1}^{\infty}\left(a_n \cos\frac{n\pi x}{l} + b_n \sin\frac{n\pi x}{l}\right), \quad (3.8.27)$$

where the coefficients have the (3.8.25) and (3.8.26) expressions.

The complex form of a Fourier series associated to a function of period $2l$ is obtained in the same way:

$$f(x) = \sum_{n=-\infty}^{+\infty} c_n e^{i\frac{n\pi x}{l}}, \quad c_n = \frac{1}{2l} \int_{-l}^{l} f(x) e^{-i\frac{n\pi x}{l}} dx, \quad n \in \mathfrak{Z}. \quad (3.8.28)$$

For even/odd function, we deduce expressions similar to formulas (3.8.23)/(3.8.24), i.e.,

$$f(x) = \frac{a_0}{2} + \sum_{n=1}^{\infty} a_n \cos\frac{n\pi x}{l},$$

$$a_n = \frac{2}{l} \int_0^l f(x) \cos\frac{n\pi x}{l} dx, \quad n \in \mathfrak{N}, \quad (3.8.29)$$

for even f and

$$f(x) = \sum_{n=1}^{\infty} b_n \sin\frac{n\pi x}{l}, \quad b_n = \frac{2}{l} \int_0^l f(x) \sin\frac{n\pi x}{l} dx, \quad (3.8.30)$$

$$n \in \mathfrak{N},$$

for odd f.

3.8.4. PERIODIC EXTENSIONS OF FUNCTIONS

A) A function given on a real interval $(a, a+2l]$, or $[a, a+2l)$, can be extended by periodicity on the real axis. Indeed, if f is defined on $(a, a+2l]$, the Ox axis is divided in intervals of length $2l$, starting from $(a, a+2l]$, to the right and to the left, and the graph of f is repeated on each interval (figure 3.11).

By this periodic extension of f, every previous considerations remains valid.

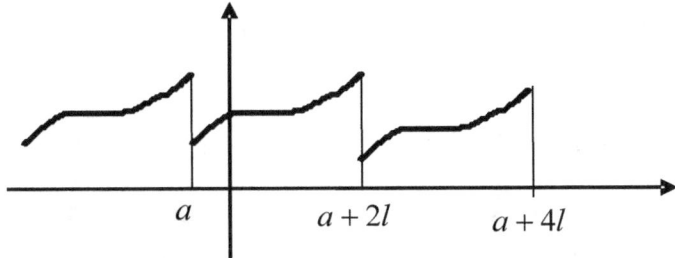

Figure 3.11. Periodic extension of a function defined on $(a, a+2l]$

B) Suppose now that f is given on a close interval – say, $[0, l]$, for simplicity. Then we can follow two steps:

1. We extend the function to an even function on the interval $[-l, l]$, taking its symmetric with respect to the Oy axis.

2. This new even function can be extended by periodicity on the Ox axis (see the steps on figure 3.12).

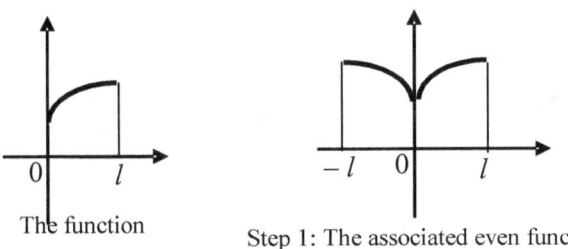

Figure 3.12. Extension by periodicity of a function defined on $[0, l]$

In this case, a *Fourier series in cosinus* is obtained.

C) A function defined on an open interval $(0, l)$ can be firstly extended, by simmetry with respect to the origin, to an odd function on $(-l, l)$. This is Step 1. Then this new odd function can be extended to a periodic odd function, just as previously (figure 3.13).

But in this case, to the obtained graph we will add the points $(kl, 0)$, $k \in \mathcal{Z}$. **The periodic odd function allows a Fourier series expansion in sinuses.**

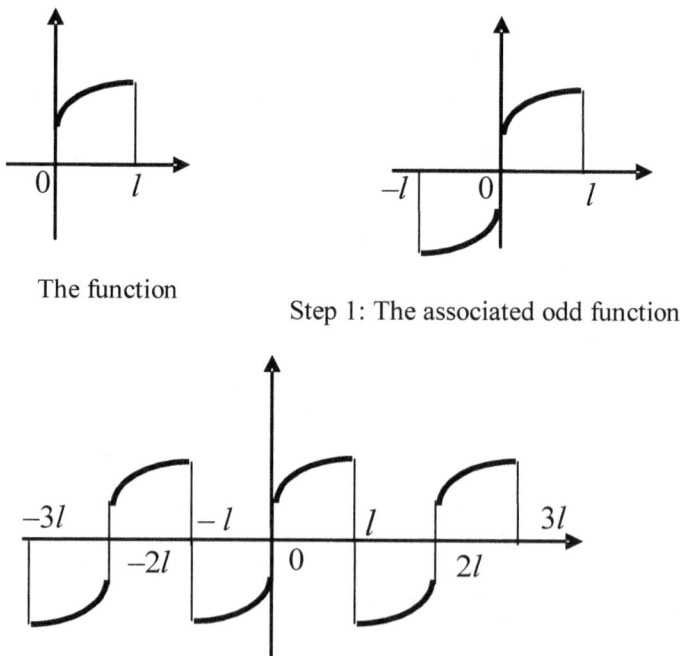

Step 2: The periodic extension

Figure 3.13. Extension by periodicity of a function defined on $(0, l)$

Remark. The values at the points kl, $k \in \mathcal{Z}$, were chosen to be null, because the arithmetic mean of the right and left limits is also null (see theorem 3.11).

EXERCISES AND PROBLEMS

1. Find the radius R of convergence and the domains D of convergence of the series

   a) $\dfrac{x+1}{1!} + \dfrac{(x+1)^2}{3!} + \dfrac{(x+1)^3}{5!} + \ldots$ A: $R = \infty$, $D = \Re$

   b) $\dfrac{x-4}{\sqrt{1}} + \dfrac{(x-4)^2}{\sqrt{2}} + \dfrac{(x-4)^3}{\sqrt{3}} + \ldots$

   A: $R = 2$, $D = [3, 5)$

   c) $\dfrac{x-1}{2} + \dfrac{(x-1)^2}{2^2} + \dfrac{(x-1)^3}{2^3} + \ldots$

   A: $R = 1$, $D = (-1, 3)$

   d) $x + (2x)^2 + (3x)^3 + (4x)^4 + \ldots$ A: conv. only at 0.

   e) $\dfrac{5x}{1!} + \dfrac{5^2 x^2}{2!} + \dfrac{5^3 x^3}{3!} + \ldots$ A: $R = \infty$, $D = \Re$

   f) $x^2 + \dfrac{x^4}{2} + \dfrac{x^6}{3} + \dfrac{x^8}{4} + \ldots$ (Hint: $x^2 = t$)

   A: $R = 1$, $D = (-1, 1)$

g) $\dfrac{x^3}{8} + \dfrac{x^6}{8^2 \cdot 5} + \dfrac{x^9}{8^3 \cdot 9} + \dfrac{x^{12}}{8^4 \cdot 13} + \ldots$

A: $R = 2, D = [-2, 2)$

h) $\dfrac{x}{2+3} + \dfrac{x^2}{2^2 + 3^2} + \dfrac{x^3}{2^3 + 3^3} + \ldots$

A: $R = 3, D = (-3, 3)$

i) $\dfrac{1}{2} \dfrac{x-1}{2} + \dfrac{2}{3} \dfrac{(x-1)^2}{2^2} + \dfrac{3}{4} \dfrac{(x-1)^3}{2^3} + \dfrac{4}{5} \dfrac{(x-1)^4}{2^4} + \ldots$

A: $R = 1, D = (-1, 3)$

j) $\dfrac{x}{1 \cdot 2} + \dfrac{x^2}{2 \cdot 3} + \dfrac{x^3}{3 \cdot 4} + \ldots$ 	A: $R = 1, D = [-1, 1]$

k) $\displaystyle\sum_{n=1}^{\infty} \dfrac{n \cdot x^n}{2^{n-1} \cdot 3^n}$ 	A: $R = 6, D = (-6, 6)$

l) $\displaystyle\sum_{n=1}^{\infty} \dfrac{(-1)^{n-1} \cdot 2^n \cdot x^n}{n}$ 	A: $R = \dfrac{1}{2}, D = \left(-\dfrac{1}{2}, \dfrac{1}{2}\right]$

m) $\displaystyle\sum_{n=1}^{\infty} \dfrac{(-2)^n - 3^n}{n} \cdot x^n$ 	A: $R = \dfrac{1}{3}, D = \left[-\dfrac{1}{3}, \dfrac{1}{3}\right)$

n) $\displaystyle\sum_{n=1}^{\infty} \dfrac{(-1)^n \cdot x^{3n+1}}{3n+1}$ 	A: $R = 1, D = (-1, 1]$

o) $\sum_{n=1}^{\infty} n^2 x^{n-1}$ A: $R = 1, D = (-1,1)$

p) $\sum_{n=1}^{\infty} \frac{2^2 + 4^2 + \ldots + (2n)^2}{1^2 + 3^2 + \ldots + (2n-1)^2} \cdot x^n$

A: $R = 1, D = (-1,1)$

q) $\sum_{n=1}^{\infty} \frac{(-1)^n \cdot (x+2)^n}{\sqrt{3n+5}}$ A: $R = 1, D = (-3,-1]$

r) $\sum_{n=1}^{\infty} \frac{(-1)^{n-1}}{n \cdot 3^n \cdot (x-5)^n}$

A: $R = 3, D = \left(-\infty, \frac{14}{3}\right) \cup \left[\frac{16}{3}, \infty\right)$

2. Find the sums of the series

a) $\frac{1}{a} + \frac{2x}{a^2} + \frac{3x^2}{a^3} + \frac{4x^3}{a^4} + \ldots$, if $|x| < a$

A: $\frac{a}{(a-x)^2}$

b) $\frac{x^2}{2a} + \frac{x^3}{3a^2} + \frac{x^4}{4a^3} + \ldots$, if $x \in [-a, a)$

A: $a \ln \frac{a}{a-x} - x$

c) $\frac{1 \cdot 2}{a^2} + \frac{2 \cdot 3}{a^3} x + \frac{3 \cdot 4}{a^4} x^2 + \ldots$, if $|x| < a$

152

$$A: \frac{2a}{(a-x)^3}$$

d) $-2x + 4x^3 - 6x^5 + 8x^7 + \ldots$, if $|x| < 1$

$$A: -\frac{2x}{(1+x^2)^2}$$

e) $\frac{2}{1}x + \frac{4}{3}x^3 + \frac{6}{5}x^5 + \ldots$ $A: \frac{x}{1-x^2} + \operatorname{arctanh} x$

Remark. We note that $\operatorname{arctanh} x = \frac{1}{2} \ln \frac{1+x}{1-x}$ for $x \in (-1, 1)$.

f) $\sum_{n=1}^{\infty} \frac{(-1)^n \cdot x^{3n+1}}{3n+1}$

$A: \frac{1}{6} \ln \frac{(1+x)^2}{x^2-x+1} + \frac{\sqrt{3}}{3} \arctan \frac{2x-1}{\sqrt{3}} + C$, $x \in (-1, 1)$

Hint: As $f(0) = 0 \Rightarrow C = \frac{\pi}{6\sqrt{3}}$

g) $\sum_{n=1}^{\infty} n^2 \cdot x^{n-1}$ $A: \frac{1+x}{(1-x)^3}$, $x \in (-1, 1)$

h) $\sum_{n=1}^{\infty} \frac{x^{4n+1}}{4n+1}$

$$A: \frac{1}{2}(\arctan x + \operatorname{arctanh} x) - x$$

i) $\sum_{n=1}^{\infty}(n+1)(n+2)x^n$ A: $\dfrac{2}{(1-x)^3}$

3. Using the method of undetermined coefficients, expand in a power series

a) $f(x)=\dfrac{1}{1-x-x^2}$

A: $1+x+2x^2+3x^3+5x^4+\ldots$

b) $f(x)=\dfrac{1}{2-3x-x^2}$

A: $\dfrac{1}{2}+\dfrac{3}{4}x+\dfrac{11}{8}x^2+\dfrac{39}{16}x^3+\ldots$

4. Using the geometric progression, expand the function $f(x)=\dfrac{1}{x}$ in a series of powers of $(x-2)$.

A: $\dfrac{1}{2}\left[1-\dfrac{x-2}{2}+\dfrac{(x-2)^2}{2^2}-\dfrac{(x-2)^3}{2^3}+\ldots\right]$

5. Write the Mac Laurin series of the functions

a) $f(x)=2^x$ A: $1+\dfrac{x\ln 2}{1!}+\dfrac{x^2\ln^2 2}{2!}+\ldots$

b) $f(x)=\sin^2 x$ A: $\dfrac{2x^2}{2!}-\dfrac{2^3 x^4}{4!}+\dfrac{2^5 x^6}{6!}-\ldots$

c) $f(x) = \ln x$, with respect to the powers of $(x-1)$

A: $(x-1) - \dfrac{(x-1)^2}{2} + \dfrac{(x-1)^3}{3} - \ldots,$

$x \in (0, 2]$

d) $f(x) = 3^x$; e) $f(x) = \cos^2 x$;

f) $f(x) = \cosh^2 x$; g) $f(x) = e^{-a^2 x^2}$;

h) $f(x) = \ln(x+a), a > 0$; i) $f(x) = e^{-x^2}$;

j) $f(x) = \ln \dfrac{\arctan x}{x}$; k) $f(x) = \ln \dfrac{\sinh x}{x}$;

l) $f(x) = \dfrac{1}{(1-x)(1+x^2)^2}$

A: $1 + x - x^2 - x^3 + 2x^4 + 2x^5 + \ldots$ (conv. for $|x| < 1$)

m) $f(x) = \ln\left(\dfrac{1+x}{1-x}\right)$, $f:(-1,1) \to \mathfrak{R}$

A: $2\sum\limits_{n=1}^{\infty} \dfrac{x^{2n-1}}{(2n-1)!}$, $(\forall) x \in (-1, 1)$

n) $f(x) = \dfrac{2x-5}{x^2 - 4x + 3}$, $f: \mathfrak{R} \setminus \{1, 3\} \to \mathfrak{R}$

A: $-\dfrac{1}{6}\sum\limits_{n=0}^{\infty}\left(9 + \dfrac{1}{3^n}\right) \cdot x^n$, $x \in (-1, 1)$

o) $f(x) = \dfrac{x}{x^2+4}$, $f: \mathfrak{R} \to \mathfrak{R}$

A: $\dfrac{x}{4} \sum\limits_{n=0}^{\infty} (-1)^n \left(\dfrac{x}{2}\right)^{2n}$, $x \in (-2, 2)$

p) $f(x) = \dfrac{x-3}{(x-1)^2}$, $f: \mathfrak{R} \setminus \{1\} \to \mathfrak{R}$

A: $-\sum\limits_{n=0}^{\infty} (2n+3)x^n$, $x \in (-1, 1)$

q) $f(x) = \dfrac{1-x}{x^2+x+1}$, $f: \mathfrak{R} \to \mathfrak{R}$

A: $(1-x)^2 \sum\limits_{n=0}^{\infty} x^{3n}$, $x \in (-1, 1)$

6. Using Taylor's formula and estimating the Lagrange remainder, approximate

a) $\sin 1°30'$ to 5 decimal places A: 0.02617

b) $\sqrt[11]{2050}$ to 4 decimal places A: 2.0001

c) $\sin 2°30'$, $\cos 2°$ to 4 decimal places

A: 0.0436, 0.9993

d) $\sqrt{e}$ to 5 decimal places A: 1.64872

e) $\sqrt[5]{1.1}$ to 4 decimal places A: 1.0192

f) $\ln 1.04$, to 4 decimal places A: 0.0392

g) $\sqrt[4]{10002}$ to 6 decimal places   A: 10.000499

h) $\cosh 0.3$ to 3 decimal places   A: 1.045

i) $\sqrt[3]{1.06}$ to 4 decimal places   A: 1.0196

j) $\sqrt[4]{260}$ to 3 decimal places   A: 4.015

k) $\ln 0.98$, to 4 decimal places   A: -0.0202

7. Using the Mac Laurin series, compute the following limits

a) $\lim\limits_{x \to 0} \dfrac{\sin x - \arctan x}{x^3}$ A: $\dfrac{1}{6}$

b) $\lim\limits_{x \to 0} \dfrac{x - \arctan x}{x^3}$ A: $\dfrac{1}{3}$

c) $\lim\limits_{x \to 0} \dfrac{1 - \cos x}{e^x - 1 - x}$ A: 1

d) $\lim\limits_{x \to 0} \dfrac{\ln\left(1 + \sin^6 x\right)}{\tan^6 x}$ A: 1

e) $\lim\limits_{x \to 0} \dfrac{e^{x^3} - 1 + \ln\left(1 + 2x^3\right)}{x^3}$ A: 3

f) $\lim\limits_{x \to 0} \dfrac{\ln(1 + 2x) - \sin 2x + 2x^2}{x^3}$ A: 4

g) $\lim\limits_{x\to\infty}\left[x - x^2 \cdot \ln\left(1+\dfrac{1}{x}\right)\right]$ \qquad A: $\dfrac{1}{2}$

Hint.: We make the substitution $x = \dfrac{1}{y}$.

h) $\lim\limits_{x\to 0} \dfrac{\sqrt[3]{1+3x} - x - 1}{1 - 4x - e^{-4x}}$ \qquad A: $\dfrac{1}{8}$

8. Using the Mac Laurin series, compute aproximately the following integrals

a) $\displaystyle\int_0^{0.5} \dfrac{1-\cos x}{x^2}\,dx$ to 4 decimal places \qquad A: 0.2483

b) $\displaystyle\int_0^{0.2} \dfrac{\sin x}{x}\,dx$ to 4 decimal places \qquad A: 0.1995

c) $\displaystyle\int_0^{0.1} \dfrac{e^x - 1}{x}\,dx$ to 4 decimal places \qquad A: 0.1025

d) $\displaystyle\int_0^{0.2} \dfrac{\ln(1+x)}{x}\,dx$ to 3 decimal places \qquad A: 0.190

e) $\displaystyle\int_0^1 e^{-x^2}\,dx$ to 3 decimal places \qquad A: 0.746

f) $\displaystyle\int_0^1 \sin\sqrt{x}\,dx$ to 3 decimal places \qquad A: 0.602

g) $\int_0^{0.5} \frac{1}{1+x^4} dx$ to 3 decimal places   A: 0.493

9. Consider the right-angle triangle ABC, whose catheti AB and AC have 1 cm., 5 cm. respectively. Using the series of $\arctan x$, $x = \frac{1}{5}$, calculate the angle opposite to the cathetus AB with a precision of 10^{-3} radians.

A: 0.197

10. Using Weierstrass' criterion, prove that the following series are uniformly convergent:

a) $\frac{1}{x^2+2} - \frac{1}{x^4+2^2} + \frac{1}{x^6+2^3} - \ldots$ on $\Re$

b) $\sin x + \frac{1}{2^2}\sin^2 2x + \frac{1}{3^2}\sin^3 3x + \ldots$ on $\Re$

c) $x + \frac{x^2}{2} + \frac{x^3}{4} + \frac{x^4}{8} + \ldots$ on $(-2, 2)$

d) $\frac{2x+1}{x+2} + \frac{1}{2}\left(\frac{2x+1}{x+2}\right)^2 + \frac{1}{4}\left(\frac{2x+1}{x+2}\right)^3 + \frac{1}{8}\left(\frac{2x+1}{x+2}\right)^4 + \ldots$, on $[-1, 1]$

e) $\frac{1}{x^2+3^0} + \frac{1}{x^2+3^1} + \frac{1}{x^2+3^2} + \ldots$ on $\Re$

11. Using Weierstrass' criterion and properties of uniformly convergent series, find out if the following series can be differentiated term by term:

a) $\arctan x + \arctan \dfrac{x}{2\sqrt{2}} + \arctan \dfrac{x}{3\sqrt{3}} + \ldots$ A: Yes

b) $(x^2+1) + 2(x^2+1)^2 + 3(x^2+1)^3 + \ldots$ A: No

c) $\dfrac{\sin 2x}{n^2 \cdot 2} + \dfrac{\sin 2^2 x}{n^2 \cdot 2^2} + \dfrac{\sin 2^3 x}{n^2 \cdot 2^3} + \ldots$ A: Yes

d) $\displaystyle\sum_{n=1}^{\infty} \dfrac{\sin\left(2^{\sqrt{n}} x\right)}{2^{\sqrt{n}} n^2}$ A: Yes

12. Using Weierstrass' criterion and properties of uniformly convergent series, find out if the following series can be integrated term by term on any finite interval $[a,b] \subset \Re$:

$$1 + \dfrac{\cos x}{1!} + \dfrac{\cos^2 x}{2!} + \dfrac{\cos^3 x}{3!} + \ldots \text{ on } \Re$$

A: Yes

13. Find the lowest positive value of x which satisfies the trigonometric equation $2\sin x - \cos x = 0$.

A: 0.4636

14. Calculate π to 2 decimals places by putting $x = \dfrac{1}{\sqrt{3}}$ in the Mac Laurin series of $\arctan x$.

A: 3.142

15. Calculate $\cos 1$ with 2 decimals places, putting $x = 1$ in the Mac Laurin development of $\cos x$.

A: 0.54

16. Expand into a Fourier series the function given on the interval $[-\pi, \pi]$ by $f(x) = \begin{cases} \pi + x, & x \in (-\pi, \pi), \\ \pi, & x \in \{-\pi, 0, \pi\}. \end{cases}$

Hint: The graph of the given function is

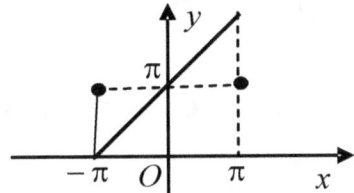

The function can be extended by periodicity to a function with period 2π over the whole real axis.

A: $f(x) = \pi + 2\sum_{n=1}^{\infty} \dfrac{(-1)^{n+1}}{n} \sin nx$

17. Expand into a Fourier series the function given on the interval $[-1, 1]$ by $f(x) = x^2$.

Hint: The graph of the function is given below:

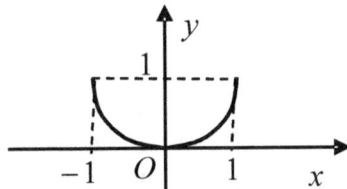

The function is even. It can be extended on $\Re$ to a periodic function of period 2.

A: $f(x) = \dfrac{1}{3} + \dfrac{4}{\pi^2} \sum\limits_{n=1}^{\infty} \dfrac{(-1)^n}{n^2} \cos n\pi x$

18. Expand into a Fourier series the function given on the interval $[0, 2]$ by $f(x) = x - \dfrac{1}{2}x^2$.

Hint: The graph of the given function is

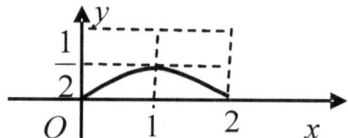

The function may be expanded in a Fourier series in various ways. We choose the two most important.

i) We first extend the function to an even one, on the interval $[-2, 2]$, then on $\Re$, to a periodic function of period 4.

A: $f(x) = \dfrac{1}{3} - \dfrac{4}{\pi^2} \sum\limits_{n=1}^{\infty} \dfrac{1+(-1)^n}{n^2} \cos \dfrac{n\pi x}{2}$

ii) We first extend the function to an odd one, on the interval $[-2, 2]$, then on $\Re$, to a periodic function of period 4.

A: $f(x) = \dfrac{8}{\pi^3} \sum\limits_{n=1}^{\infty} \dfrac{1-(-1)^n}{n^3} \sin \dfrac{n\pi x}{2}$

19. Expand into a Fourier series the function given on the interval $[-l, l)$ by

$$f(x) = \begin{cases} 0, & -l \le x < 0, \\ x, & 0 \le x < \dfrac{l}{2}, \\ \dfrac{l}{2}, & \dfrac{l}{2} \le x < l. \end{cases}$$

Hint: The graph of the given function is

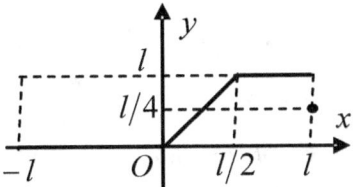

The function can be extended on $\Re$ to a periodic function of period $2l$.

A: $f(x) = \dfrac{3l}{16} + l\left(-\dfrac{1}{\pi^2}\cos\dfrac{\pi x}{l} + \dfrac{2+\pi}{2\pi^2}\sin\dfrac{\pi x}{l}\right) +$

$+ l\left(-\dfrac{1}{2\pi^2}\cos\dfrac{2\pi x}{l} - \dfrac{1}{4\pi}\sin\dfrac{2\pi x}{l}\right) +$

$+ l\left(-\dfrac{1}{9\pi^2}\cos\dfrac{3\pi x}{l} + \dfrac{-2+3\pi}{18\pi^2}\sin\dfrac{3\pi x}{l}\right) + \ldots$

20. Expand into a Fourier series the function given on the interval $(-\pi, \pi)$ by $f(x) = x$.

Hint: It is an odd function. It can be extended on $\Re$ to an odd periodic function f^* of period 2π, also adding the values $f^*((2k+1)\pi) = 0$, $k \in \mathcal{Z}$.

A: $f(x) = -2\sum\limits_{n=1}^{\infty}(-1)^n \dfrac{\sin nx}{n}$

21. Expand into a Fourier series the function given on the interval $(-\pi, \pi)$ by $f(x) = x^3$.

   Hint: It is an odd function. It can be extended on $\mathfrak{R}$ to an odd periodic function f^* of period 2π, also taking $f^*((2k+1)\pi) = 0$, $k \in \mathfrak{Z}$.

   A: $f(x) = \sum_{n=1}^{\infty} (-1)^n \left(\frac{12}{n^3} - \frac{2\pi^2}{n} \right) \sin nx$

22. Expand into a Fourier series the function given on the interval $[-1, 1]$ by $f(x) = |x|$.

   Hint: It is an even function. It can be extended on $\mathfrak{R}$ to an even periodic function f^* of period 2.

   A: $f(x) = \frac{1}{2} - \frac{4}{\pi^2} \sum_{n=1}^{\infty} \frac{\cos(2n+1)\pi x}{(2n+1)^2}$

23. Expand into a Fourier series the function given on the interval $(0, \pi)$ by $f(x) = \pi - 2x$, extending it

   i) to an odd function on $(-\pi, \pi)$

   A: $f(x) = 2 \sum_{n=1}^{\infty} \frac{\sin 2nx}{n}$

   ii) to an even function on $[-\pi, \pi]$

   A: $f(x) = \frac{8}{\pi} \sum_{n=0}^{\infty} \frac{\cos(2n+1)\pi x}{(2n+1)^2}$

24. Expand into a Fourier series of sines the function given on the interval $(0,1)$ by $f(x) = x$.

A: $f(x) = \dfrac{2}{\pi} \sum\limits_{n=1}^{\infty} (-1)^{n+1} \dfrac{\sin n\pi x}{n}$

25. Expand into a Fourier series of cosines the function given on the interval $[0,2]$ by

$$f(x) = \begin{cases} x, & x \in [0,1), \\ 2-x, & x \in (1,2]. \end{cases}$$

Hint: It is an even function. It can be extended on $\Re$ to an even periodic function f^* of period 2.

A: $f(x) = \dfrac{1}{2} - \dfrac{4}{\pi^2} \sum\limits_{n=0}^{\infty} \dfrac{\cos(2n+1)\pi x}{(2n+1)^2}$

Remark. This function can also be expanded in a Fourier series of sines, first extending it it to an odd function on the interval $[-2,2]$, and then to a periodic function of period 4 on $\Re$.

A: $f(x) = \dfrac{8}{\pi^2} \sum\limits_{n=1}^{\infty} \dfrac{(-1)^{n+1}}{(2n-1)^2} \sin \dfrac{(2n-1)\pi x}{2}$

REFERENCES

1. BÂRZĂ, I., *Analiză Matematică. Culegere de Probleme Rezolvate* (Mathematical Analysis. A collection of solved problems), Niculescu Printing House, Bucharest, 2002.
2. CIORĂNESCU, N., *Curs de Algebră şi Analiză Matematică* (Course of Algebra and Analysis), Ed. Tehnică, Bucharest, 1958.
3. CRAW, I., *Advanced Calculus and Analysis*, Univ. of Aberdeen, 2000.
4. COURANT, R., *Differential & Integral Calculus*, t.2, Blackie and Son Ltd, London and Glasgow, 1936.
5. DANKO, P.E., POPOV, A.G., *Vîsşaia Matematika v uprajneniah i zadachah*, Vîsşaia Şkola, Moscva, 1964 (Advances mathematics in problems and exercises).
6. KÖRNER, T.W., *Fourier Analysis*, Cambridge University Press, 1988, http://www.cambridge.org
7. NIŢĂ, A., STĂNĂŞILĂ, T., *1000 de probleme rezolvate şi exerciţii fundamentale pentru studenţi şi elevi* (1000 solved problems and fundamental exercices for students), Ed. BIC ALL, Bucharest, 1997.
8. TOMA, I., *Analyse Mathématique. Calcul différentiel*, (Mathematical Analysis. Differential Calculus) Conspress, Bucharest, 2010.
9. TOMA, I., MOŞNEGUŢU, V., CONSTANTINESCU, Şt., *Analyse Mathématique. Équations différentielles ordinaires. Calcul intégral*, Ed. Conspress, Bucharest, 2014.

10. TOMA, I., MOŞNEGUŢU, V., CONSTANTINESCU, Şt., *Ordinary Differential Equations*, CreateSpace Independent Publishing Platform, 2016, ISBN - 13: 978-1540318015, ISBN - 10: 154031801X.

11. TOMA, I., MOSNEGUŢU, V., CONSTANTINESCU, Şt., *Integral Calculus,* CreateSpace Independent Publishing Platform, 2017, ISBN-13: 978-1548789909, ISBN-10:1548789909.

Links:

12. DAWKINS, P., *Calculus II, Sequences and Series*, 2007, http://tutorial.math.lamar.edu/terms.aspx, http://www.csun.edu/matabots/resources/CalcII_Seq_Series.pdf

13. FOWLER, J., SNAPP, B., *Sequences and Series*, Creative Commons license, 2014,
https://archive.org/details/mooculus-sequences-book

14. GREEN, J. A., *Sequences and Series*, Routledge & Kegan Paul Ltd., 1958,
https://archive.org/details/SequencesAndSeries

15. KAYE, R., *Sequences and Series*, Gnu Free Documentation Licence, 2006/2015
http://www.freebookcentre.net/maths-books-download/Sequences-and-Series.html

16. HUTCHINSON, J. E., *Introduction To Mathematical Analysis,* 1994,
https://maths- people.anu.edu.au/~john/Assets/Lecture Notes/B21H_97.pdf

17. RUDIN, W., *Principles of Mathematical Analysis*, McGraw-Hill, Inc., 1976, 3rd ed.,

https://notendur.hi.is/vae11/Pekking/principles_of_mathematical_analysis_walter_rudin.pdf

18. TRENCH, W.F., *Introduction to real analysis*, http://ramanujan.math.trinity.edu/wtrench/texts/TRENCH_REAL_ANALYSIS.PDF

19. JIRKA, L., *Basic Analysis. Introduction to real analysis*, 2016, http://www.jirka.org/ra/realanal.pdf

20. THOMSON, B. S., BRUCKNER, J. B., BRUCKNER, A. M., *Real Analysis*, second edition, 2008, http://classicalrealanalysis.info/com/documents/BBTAlllChapters-Landscape.pdf